MYCOTOXINS

MYCOTOXINS

Formation, Analysis and Significance

J. E. Smith
Division of Applied Microbiology
Department of Bioscience and Biotechnology,
University of Strathclyde, Glasgow

and

M. O. Moss
Department of Microbiology,
University of Surrey, Guildford

JOHN WILEY & SONS
Chichester · New York · Brisbane · Toronto · Singapore

Library of Congress Cataloging in Publication Data:

Smith, John E.
Mycotoxins : formation, analysis, and significance.
Includes index.
Mycotoxins. I. Moss, Maurice O.
II. Title.
QP632.M9S63 1985 615.9'5292 84-26953
ISBN 0 471 90671 9

British Library Cataloguing in Publication Data:

Smith, John E. (John Edward)
Mycotoxins : formation, analysis and significance.
1. Mycotoxins
I. Title II. Moss, Maurice O.
574.2'326 QP632.M9

ISBN 0 471 90671 9

Printed in Great Britain

Contents

CHAPTER 1

Introduction

WHAT ARE FUNGI, TOADSTOOLS AND MOULDS?

The fungi is a trivial name given to a very broad grouping of living organisms. They are eukaryotes, that is to say, they have well-defined membrane-bound nuclei with a number of chromosomes, and so are clearly distinguishable from bacteria. They are heterotrophic, requiring organic carbon compounds of varying degrees of complexity, which distinguishes them from plants. All but a few fungi have a well-defined cell wall, through which all their nutrients have to pass in a soluble form, and, in this respect, they differ from animals.

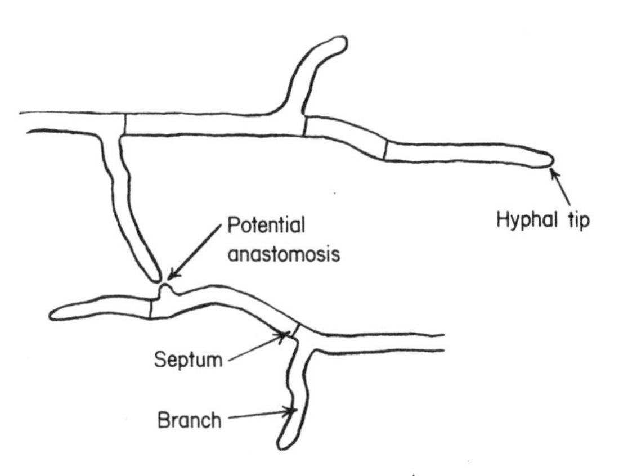

Fig. 1.1. Hyphal growth and differentiation

This broad assemblage includes single-celled organisms, aquatic organisms, animal and plant parasites, but it is those which have become adapted to growth over, and through, solid substrates encountered in the terrestrial environment which will concern us most. They effectively achieve this by growing as filaments, or hyphae, which extend at the tips, branching and anastomosing to form a complex mycelium (Fig. 1.1). Although a mycelium may grow quite extensively, there are limitations once the substrate has been exhausted and the need arises to disperse to new sources of nutrients. Filamentous growth is thus closely associated with the production of spores, the majority of which are disseminated through the atmosphere, although some may be spread via water or biological agents such as insects.

The requirement to produce large numbers of spores has led to the elaboration of complex macroscopic fruiting bodies by many of the Basidiomycetes and some of the Ascomycetes. Complex though these toadstools, mushrooms and morels may be, careful dissection and study with the microscope will reveal the fundamentally filamentous architecture of branching, anastomosing hyphae (Fig. 1.2).

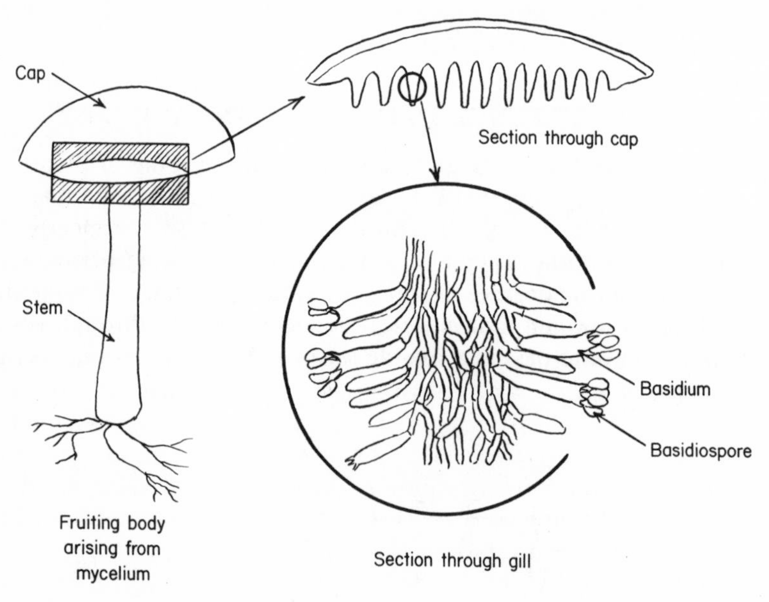

Fig. 1.2. Formation and function of toadstools and mushrooms

MYCETISMS

The desirability of certain species of agarics and morels as food was clearly recorded in the literature of the Romans and Greeks as was the knowledge that some are deadly poisonous. Equally ancient are some of the misconceptions about how to distinguish the poisonous and edible species. Thus, Horace stated that all the toadstools that live in meadows are good to eat and Dioscorides claimed that poisonous toadstools have a thick layer of mucilage and decay very quickly after being collected.

Today *Agaricus bisporus*, and closely related species, are cultivated throughout the world to form a multimillion pound industry. The harvesting of wild species is also widespread but, despite the availability of information about poisonous toadstools, illness and death still occur as a result of consuming poisonous species such as *Amanita phalloides* and

Cortinarius orellanus. Many of the toxins of these fungi are well characterized (Table 1.1) but they are not usually considered as mycotoxins and will not be dealt with in this book. Such poisonings following the deliberate or unwitting consumption of these larger fungi are conveniently referred to as mycetisms.

Table 1.1. Examples of poisonous toadstools and their toxins

Species	Common name	Principal toxin
Amanita phalloides	Deathcap	Phalloidin
Inocybe patouillardii	Red-staining inocybe	Muscarine
Cortinarius orellanus	Mountain cortinarius	Orellanine
Gyromytra esculenta[a]	False morel	*N*-methyl, *N*-formyl hydrazine

[a] This Ascomycete is edible after thorough washing and cooking to remove the toxin.

MYCOTOXINS AND MYCOTOXICOSES

Amongst the filamentous fungi there are many which do not produce macroscopic fruiting bodies but are nonetheless important because of their ability to grow on foods, animal feeds or the raw materials used in the manufacture of these commodities. Commonly referred to as the moulds many of these fungi are able to produce a wide range of secondary metabolites (see Chapter 3). Some of these metabolites are pigments, some have antibiotic properties and some are toxic to plants and animals. The mycotoxins are those mould metabolites which cause illness or death of man, or his domesticated animals, following consumption of a contaminated food. The illness itself is referred to as a mycotoxicosis.

It should be noted that there is an anomaly in the use of the term mycotoxin. Phytotoxins and zootoxins are compounds toxic to plants and animals respectively, and by analogy the term mycotoxin might be expected to refer to compounds toxic to fungi but it is used to describe compounds produced by fungi toxic to animals in the restricted sense outlined above.

For farm animals, or human populations whose diet is made up largely of plant products, mycotoxins may be produced directly by growth of moulds on animal feed or human food. Illnesses arising from eating such food are conveniently referred to as primary mycotoxicoses. It is also possible that mycotoxins may pass through the food chain into animal products, such as milk or meat, which have not themselves been contaminated by mould growth. Illnesses arising from such sources are referred to as secondary mycotoxicoses (Fig. 1.3).

Although mycotoxins are quite clearly defined in this pragmatic manner, they are a very diverse group of compounds, produced by a taxonomically wide range of filamentous fungi (Chapter 2), and showing a diverse range of toxic effects (Fig. 1.4).

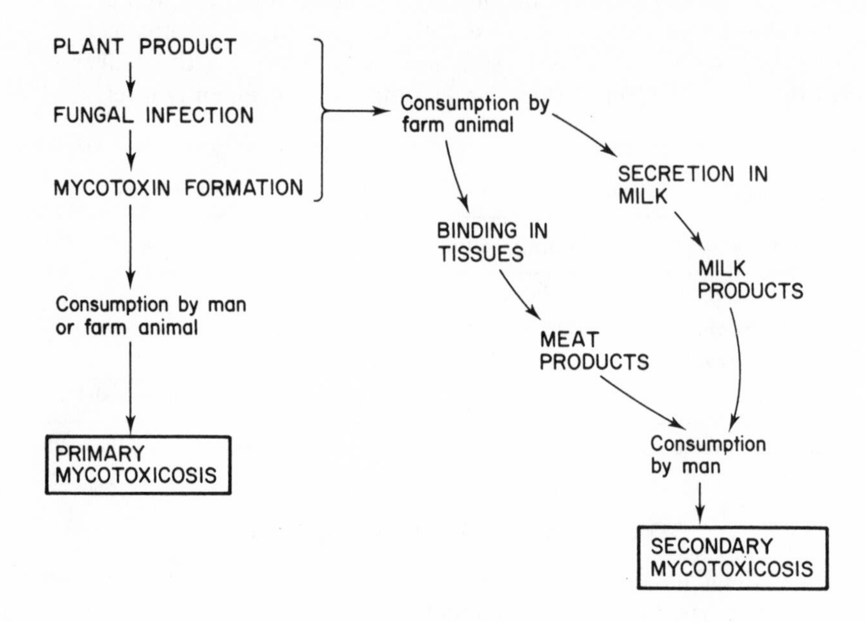

Fig. 1.3. Primary and secondary mycotoxicoses

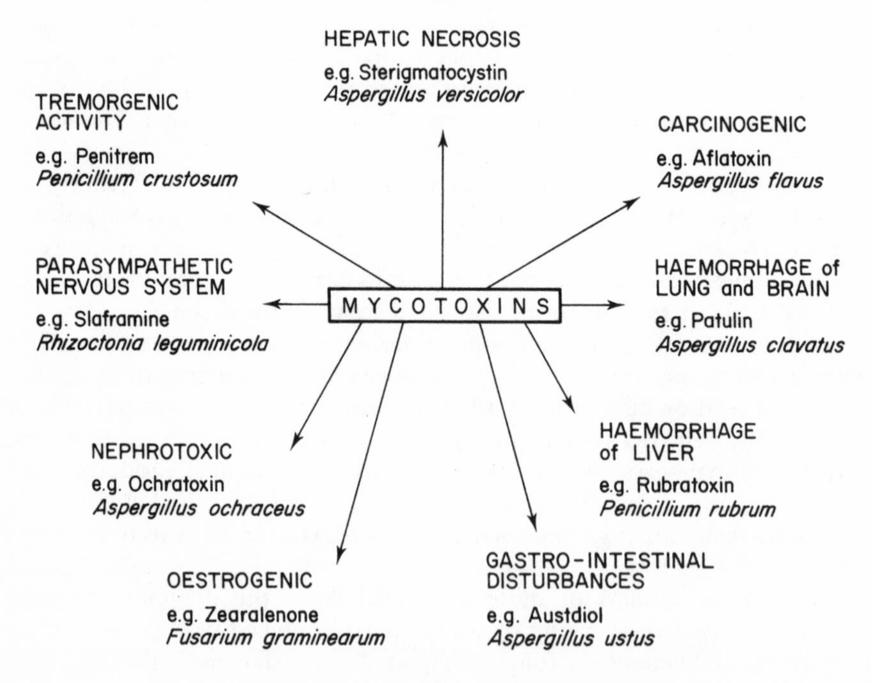

Fig. 1.4. Toxic phenomena associated with mycotoxins

ERGOTISM: AN HISTORIC MYCOTOXICOSIS

There are records of outbreaks of gangrenous ergotism, occurring during the ninth and tenth centuries, in which there are vivid descriptions of limbs rotting and falling off. These outbreaks became particularly frequent in France and, during the eleventh century, the Order of St Anthony was founded to provide hospitals for those suffering with what came to be known as St Anthony's fire. The disease is now known to be associated with the consumption of cereals, particularly rye, contaminated with the sclerotia, or resting structures, of the plant pathogenic fungus *Claviceps purpurea* (Fig. 1.5).

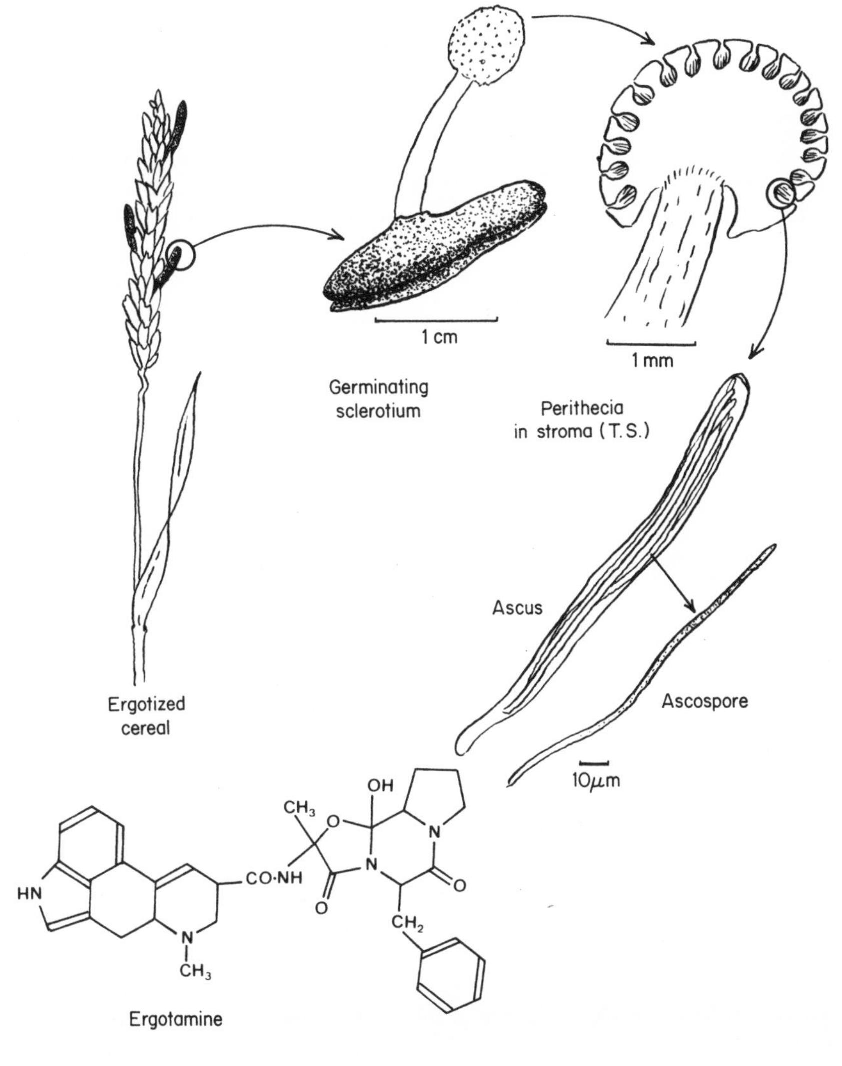

Fig. 1.5. *Claviceps purpurea* and its toxins

Another form of ergotism affects the nervous system causing convulsions and tonic spasms of the limbs. A symptom during the early stages of poisoning by this fungus is vividly described as the sensation of ants running about beneath the skin. This second type of mycotoxicosis, although involving the same fungus, was more common in Russia and Eastern Europe and the differences in symptoms were probably due to a multifactorial interaction of diet, dose and the production of different mixtures of alkaloids.

Although tens of thousands of people died from these awful intoxications they are now rare amongst humans, but outbreaks occur occasionally in farm animals, and the ergot alkaloids have found valuable uses in medicine.

AFLATOXIN: AN INTERNATIONAL MYCOTOXIN

In different parts of the world there have been outbreaks of mycotoxicosis which were studied intensively at a national level and will be dealt with in later chapters. Thus, in man, alimentary toxic aleukia has been a severe problem in parts of Russia, and yellow rice disease caused grave concern in Japan. Amongst animals, sheep facial eczema has been the subject of considerable research in New Zealand, while red clover disease, vulvovaginitis and mouldy corn toxicosis were all initially studied in the United States.

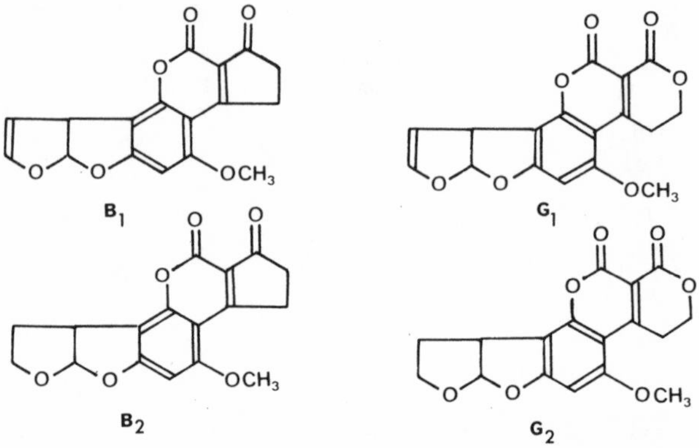

Fig. 1.6. The aflatoxins

However, in 1959, a rather singular event occurred in Britain which resulted in the study of mycotoxins becoming truly international. A turkey farm in East Anglia suffered dramatic losses of thousands of turkey poults over a very short period of time. It was quickly demonstrated that these birds died from a poison present in the pelleted feed which formed a major part of their diet. Two industries were threatened by this event: the intensive turkey-rearing industry producing birds for the Christmas table, and the pelleted feed industry which supported it. The major raw materials used by the latter were protein-rich plant products such as groundnut meal, imported from tropical and subtropical countries, the economies of which were frequently supported by the export of such commodities.

Examination of the incriminated groundnut meal revealed the presence of mould mycelium and thin layer chromatography showed the presence of several new compounds which fluoresce intensively under ultraviolet light. The mould was shown to be *Aspergillus flavus* and its highly toxic metabolites have been called the aflatoxins (Fig. 1.6).

There is no question about the acute toxicity of the aflatoxins (Table 1.2) and it is certain that they have caused the deaths of human beings, but it was the demonstration that these compounds are potent carcinogens in some animal species which triggered off an intensive international research effort resulting in a massive literature on this single group of mycotoxins. In the early 1960s the terms mycotoxin and aflatoxin were synonomous!

Table 1.2. Acute toxicity of aflatoxin B_1 expressed as a single oral dose LD_{50} [a]

Species	LD_{50} mg kg^{-1} bodyweight
Rabbit	0.3
Duckling (1 day old)	0.34
Cat	0.55
Pig	0.6
Rainbow trout	0.8
Dog	0.5–1.0
Sheep	1.0–2.0
Guinea pig	1.4–2.0
Baboon	2.0
Chicken	6.3
Rat (male)	5.5–7.2
Rat (female)	17.9
Macaque (female)	7.8
Mouse	9.0
Hamster	10.2

[a] The LD_{50} is the dose that will cause the death of 50 per cent of a statistically significant population. It is usually obtained by extrapolation from dose response experiments.

Once the chemical and biological nature of the aflatoxins had been established it became evident that many instances of aflatoxin poisoning had, in fact, been observed previously in both laboratory animals and intensively reared animals. At the time these diseases had to be documented as being of unknown aetiology. One such disease was liver carcinoma in rainbow trout which was associated with the introduction of cottonseed meal to replace slaughterhouse offal as the protein component of pelleted feeds used in fish farms. At the time the carcinogen could not be identified although it was suspected that natural constituents of cottonseed meal, such as cycloprenoid fatty acids, could be acting as co-carcinogens. Once the carcinogenic properties of the aflatoxins had been recognized it was shown that this same group of compounds were present in minute concentrations in cottonseed meal and that the rainbow trout is extremely sensitive to aflatoxin B_1. Pelleted feeds containing as little as 0.4 $\mu g\ kg^{-1}$ can cause a significant incidence of hepatoma.

ECONOMIC, SOCIAL AND POLITICAL IMPACTS OF MYCOTOXINS

In some situations it is possible to make a fairly direct assessment of economic losses due to the presence of mycotoxins in agricultural commodities. Thus, facial eczema of sheep, caused by sporidesmins produced by *Pithomyces chartarum* (see Chapter 4), cost New Zealand farmers an estimated $100 million in 1981. Aflatoxin contamination of maize in the South Eastern states of the USA in 1977 cost farmers and handlers an estimated $111 million. The presence of deoxynivalenol, produced by *Fusarium graminearum* and other fusaria growing on winter wheat, has cost the Canadian farmers several million dollars during the last few years.

But, how does one assess the economic loss following the deaths of several hundred people in an Indian village after eating moulded food contaminated with aflatoxin? Indeed, in general, the economic losses resulting from mycotoxin-contaminated foods and feeds can be recognized qualitatively but are usually difficult to quantify. Losses may arise from the need to dispose of a contaminated commodity or from the additional costs of detoxification. There may be losses from the deaths of livestock, lower growth rates or decreased feed efficiency. Failure to comply with stringent controls imposed by one country will result in the loss of export markets. In Third World Countries, efforts to maintain export markets may lead to the consumption of poorer quality materials by the indigenous population with an increased exposure to mycotoxins. A wealthy nation can impose an unjustifiably stringent control to protect is own farming community from the economic pressures of cheaper imports from poorer countries.

These are just some of the interactions through which mycotoxins may have an impact on the economics and politics of trading nations.

FURTHER READING

Anon (1979). Mycotoxins, *Environmental Health Criteria*, **11**. World Health Organisation, Geneva.

Goldblatt, L. A. (Ed.) (1969). *Aflatoxin: Scientific Background, Control and Implications*, Academic Press, New York.

Heathcote, J. G. and Hibbert, J. R. (1978). *Aflatoxins: Chemical and Biological Aspects*, Elsevier Scientific, Amsterdam.
Mantle, P. G. (1977). Chemistry of *Claviceps* mycotoxin. In: *Mycotoxigenic Fungi, Mycotoxins, Mycotoxicoses: an Encyclopaedic Handbook*, T. D. Wyllie and L. G. Morehouse, (Eds) Marcel Dekker, New York, pp. 421–426.
Moreau, C. (1978). *Larouse des Champignons*, Librairie Larousse, Paris.
Moreau, C. (1979). *Moulds, Toxins and Food*, Wiley, Chichester.
Ramsbottom, J. (1953). *Mushrooms and Toadstools, a Study of the Activities of Fungi*, Collins, London.
Sargeant, K., Sheridan, A., O'Kelly, J. and Carnaghan, R. B. A. (1961). Toxicity associated with certain samples of groundnuts. *Nature (London)*, **192**, 1096–1097.

CHAPTER 2

The Toxigenic Fungi

MICROFUNGI: YEASTS AND MOULDS

Within the broad description of the fungal kingdom outlined in Chapter 1, there are a number of groups, the chytrids, hyphochytrids and the oomycetes, which are primarily associated with aquatic habitats. Many of them have life cycles which include motile stages capable of swimming in water. These aquatic fungi are not known to produce toxins and the mycotoxicologist is mainly concerned with those fungi which have adapted to terrestrial habitats. Of the terrestrial fungi, it is the filamentous microfungi, commonly called moulds, which produce mycotoxins. The single-celled fungi, or yeasts, are not known to produce mycotoxins although they are of considerable interest to food and industrial mycologists for other reasons. The yeasts and yeast-like fungi can be isolated from soil, plant and animal secretions and are probably a very specialized group of fungi. Their relationship with the rest of the fungi is discussed in detail by Von Arx (1979).

TOXIGENIC MOULDS

The filamentous mycelium of the moulds is well adapted for growth over surfaces and through solid substrates. They can secrete enzymes to break down and absorb the available nutrients so that energy, active cytoplasm and materials can be transported to the actively growing hyphal tip. This structure of long, thin, branching tubes provides a large surface area relative to the biomass which it supports and it is through the cell walls that secondary metabolites, such as mycotoxins, may be secreted into the surrounding substrate.

The production of toxic metabolites is not confined to a single group of moulds irrespective of whether they are grouped according to structure, ecology or phylogenetic relationships. Fig. 2.1 shows a widely accepted scheme for classifying the fungi and Table 2.1 gives an indication of the classes to which some of the genera of interest to food mycologists belong. The nomenclature used is that found in the most recent edition of *Ainsworth and Bisby's Dictionary of the Fungi* (Hawksworth *et al.*, 1983), a valuable source of information about the fungi.

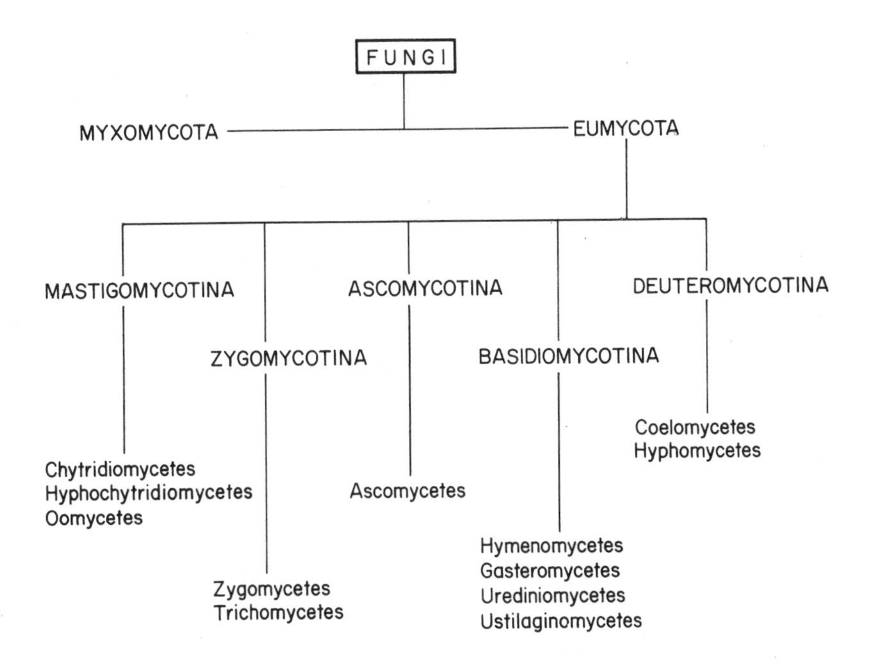

Fig. 2.1. The relationships of the higher categories of fungi proposed in Hawksworth *et al*. (1983)

OOMYCETES AND ZYGOMYCETES

These two classes have, in the past, been considered together and referred to as Phycomycetes but they are now placed in different subdivisions. Although the Oomycetes are essentially aquatic fungi some, such as the Perenosporales, have adapted to a more terrestrial habitat as pathogens of plants. Indeed, they have evolved the ability to produce airborne sporangia (Fig. 2.2) for dispersal. Of the many species of downy mildew diseases of plants, *Phytophthora infestans* is the best known as the agent of potato blight. Although there is evidence that it produces high molecular weight phytotoxins, and there has been some speculation that it may play a role in congenital disorders such as spina bifida, this and other downy mildews have not been implicated in any known mycotoxicoses.

Amongst the Zygomycetes it is the Mucorales which have the most impact in food mycology. They are clearly associated with terrestrial habitats, and the evolution of asexual sporangiospores has adapted them for aerial and insect dispersal (Fig. 2.3), although they do require high water activities for active growth. The order includes saprophytes, some of which are important in food spoilage, as well as plant pathogens. Some species of *Mucor*, *Absidia* and *Rhizopus* have been implicated in mycotic diseases of animals. Although there are occasional reports of ability to produce toxic metabolites there is no strong evidence to implicate this group of fungi in mycotoxicoses.

Table 2.1. Major groups of terrestrial filamentous fungi

Class	Order	Examples of genera	Comments
Oomycetes	Peronosporales	*Phytophthora*	Plant pathogens including potato blight. Not known to produce mycotoxins
Zygomycetes	Mucorales	*Mucor* *Rhizopus*	Important as agents of food spoilage. Occasional reports of mycotoxins
Ascomycetes	Clavicipitales	*Claviceps* *Eurotium*	Plant pathogens, ergotism Saprophytes able to grow at low a_w. Food spoilage some toxigenic
	Hypocreales	*Nectria* *Gibberella*	Plant pathogens, some toxigenic.
	Sphaeriales	*Chaetomium*	Saprophytes, toxigenic
	Pezizales	*Helvella* *Gyromitra*	Poisonous 'toadstools'
Ustilaginomycetes	Ustilaginales	*Ustilago*	Plant pathogens, smuts
Urediniomycetes	Uredinales	*Puccinia*	Plant pathogens, rusts
Hymenomycetes	Agaricales	*Amanita* *Agaricus*	Poisonous and edible 'toadstools', mushrooms
Coelomycetes	–	*Phomopsis*	Pycnidial 'fungi imperfecti'
Hyphomycetes	–	*Aspergillus* *Penicillium* *Fusarium* *Pithomyces* *Alternaria* *Stachybotrys*	The 'fungi imperfecti' including many toxigenic species (Table 2.3)
Mycelia Sterilia	–	*Rhizoctonia*	Plant pathogens including at least one toxigenic species

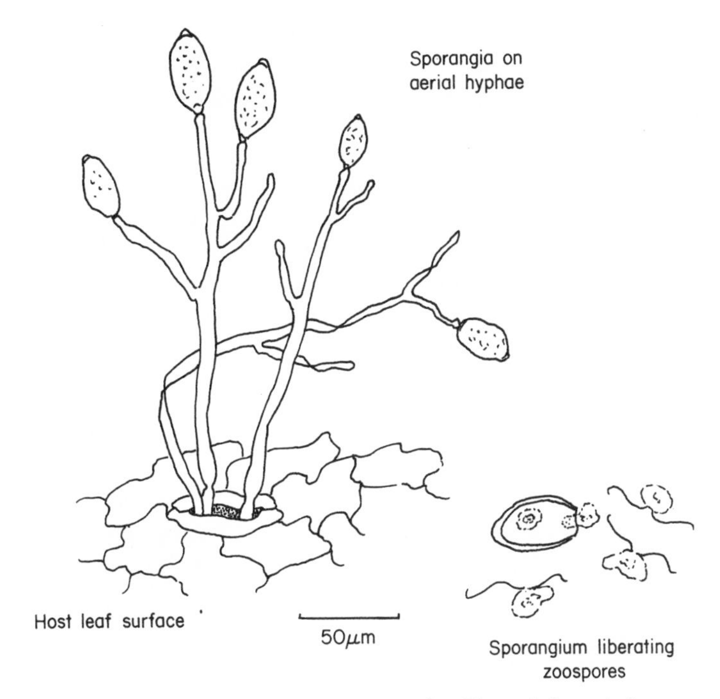

Fig. 2.2. Sporangium development in *Phytophthora infestans*

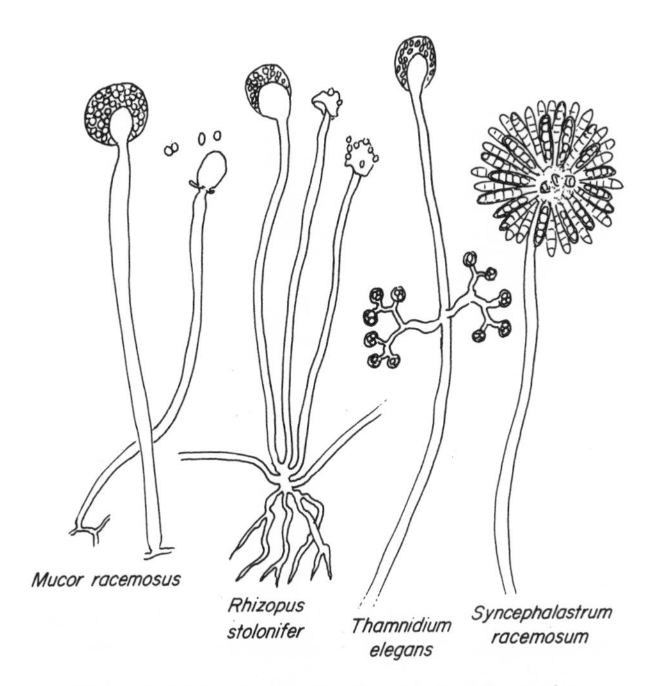

Fig. 2.3. Asexual reproduction in the Mucorales

ASCOMYCETES

This class contains a diverse assemblage of fungi all characterized by the production of ascospores within an ascus as a result of sexual reproduction (Fig. 2.4). The ascus may be naked (Hemiascomycetes), enclosed in groups in a partial or complete structure known as a cleistothecium (Plectomycetes), enclosed in a flask-shaped structure referred to as a perithecium (Pyrenomycetes), or form part of the surface tissue of a more or less cup-shaped structure (Discomycetes) which may close in on itself to form the underground fruiting body of the truffles (Tuberales). A somewhat distinct group of Ascomyetes with double-walled (bitunicate) asci are grouped together as the Loculoascomycetes.

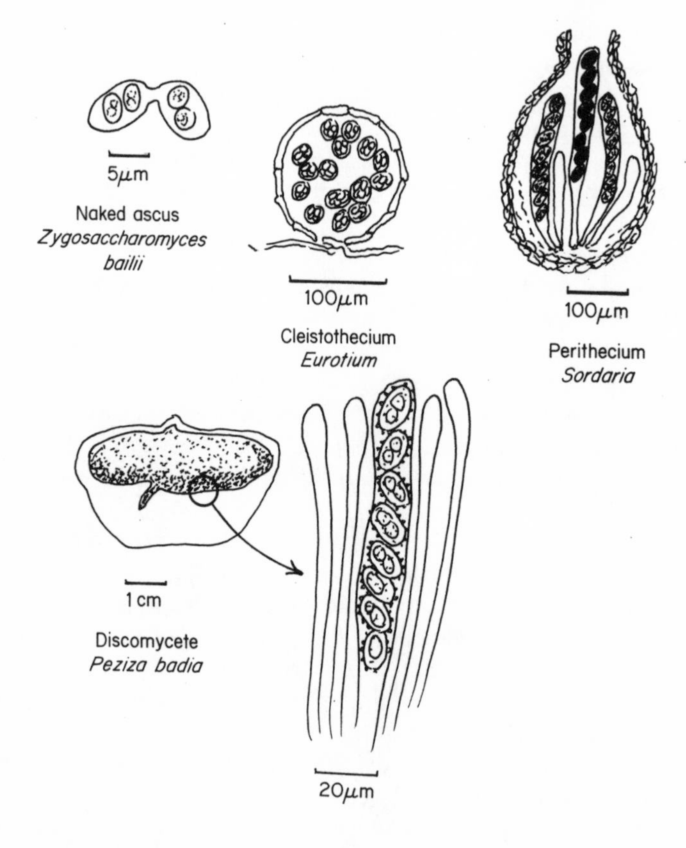

Fig. 2.4. Diversity of structure in fruiting bodies of Ascomycetes

For terrestrial fungi, dispersal from an exhausted nutrient base to new substrates is an important aspect of their biology. In some groups, such as coprophilous Ascomycetes and many plant pathogens, ascospores are actively discharged and provide a very effective agent of dispersal. In other groups, the role of dispersal is carried out by spores produced asexually, conidiospores, which will be discussed in more detail in the section on the Deuteromycotina.

Although they may not always be used for dispersal, the sexual process itself may be important in the evolution of new genetic combinations, and ascospores may have enhanced resistance to adverse environmental conditions. Thus the ascospores of the species of *Byssochlamys* are sufficiently heat-resistant to survive the heat processing of canned fruits and these fungi may be a nuisance causing the spoilage of such commodities. Isolates of *Byssochlamys*, and their conidial stages which are placed in the anamorph genus *Paecilomyces*, can produce a number of metabolites of varying toxicity, including byssochlamic acid, byssotoxin A, variotin and patulin, but there is no evidence of their production in canned fruit.

Byssochlamys belongs to the order Eurotiales which also includes such genera as *Eurotium* and *Eupenicillium* with imperfect states (anamorphs) in the genera *Aspergillus* and *Penicillium*, respectively. These latter are very important in the context of mycotoxins and will be discussed in more detail later. The use of different names for different stages of the same organism seems unnecessarily confusing but it is something which mycologists learn to live with and often find convenient. A genus like *Aspergillus* may contain the anamorphic states of species from several genera of Ascomycetes. Amongst the Hypocreales there are genera such as *Nectria* and *Gibberella*, some species of which have anamorphs in the genus *Fusarium* which must also be considered in more detail because of its importance in mycotoxicology. The genus *Nectria* is an example of a teleomorph, or perfect state genus, which includes species with anamorphic states in a number of genera. As well as *Fusarium* there are *Dendrodochium*, *Myrothecium*, *Cylindrocarpon* and *Gliocladium*, all of which are associated with mycotoxin production.

Whereas the Hypocreales often produce brightly coloured fleshy perithecia, the Sphaeriales produce dark frequently carbonaceous perithecia. A representative genus of the Sphaeriales is *Chaetomium*, species of which are frequently cellulolytic and associated with the decay of senescent plant material. There have been reports of materials contaminated with *Chaetomium globosum* being toxic to animals, and a number of toxic metabolites have been isolated and characterized in the laboratory. However, such reports have not been widespread and the significance of *Chaetomium* as a toxigenic genus still has to be assessed.

An important Ascomycete, discussed in Chapter 1, is *Claviceps purpurea*. It is a parasite of grasses and cereals and is closely related to a number of genera with fascinating life histories. *Cordyceps* includes species parasitic on insects, as well as species parasitic on truffles, whereas *Epichloe* includes species causing choke disease of grasses as well as others which may have become highly specialized as benign endophytes of grasses. One such endophyte, the taxonomic position of which is still uncertain, may be directly or indirectly responsible for the presence of a group of neurotoxic metabolites, known as lolitrems, in perennial ryegrass. The lolitrems, which may be metabolites of the endophyte or plant metabolites produced in response to the presence of the endophyte, are thought to be responsible for perennial ryegrass staggers disease (Gallagher *et al.*, 1981).

Table 2.2 summarizes the relationships between some Ascomycetes of interest in the context of mycotoxin formation and their anamorphs.

Table 2.2. Ascomycetes and their anamorphs associated with mycotoxin formation

Teleomorph	Anamorph	Mycotoxins
Claviceps purpurea	*Sphacelia segetum*	Ergot alkaloids
Eurotium chevalieri	*Aspergillus chevalieri*	Xanthocillin
Eupenicillium ochrosalmoneum	*Penicillium ochrosalmoneum*	Citreoviridin
Monographella nivalis	*Fusarium nivale* [a]	Trichothecenes
Gibberella zeae	*Fusarium graminearum*	Zearalenone Trichothecenes
Nectria haematococca	*Fusarium solani*	Trichothecenes
Hypocrea spp.	*Trichoderma viride*	Trichodermin

[a]Also referred to as *Gerlachia nivalis.*

DEUTEROMYCETES

Sexual reproduction by many filamentous fungi frequently requires the physical anastomosis, or fusion, of two compatible strains. The ascospores or basidiospores produced by sexual reproduction are haploid and the mycelium which develops following germination of each spore is also haploid. There has been a strong evolutionary pressure on each strain to develop mechanisms for successful dispersal which are independent of sexual reproduction. This has led to the formation of a diverse range of processes for packaging small fragments of the haploid mycelium as conidiospores for dispersal. In many instances this process has been so successful that the sexually compatible strains may have become geographically removed or there may be other reasons why sexual reproduction no longer occurs, or is rare. We then isolate and observe these fungi as the anamorphs (= imperfect, asexual state). Mycologists have attempted to classify them and name them, a universally accepted name being particularly important when identifying isolates from many parts of the world.

Some of the structures associated with the formation of conidiospores by Deuteromycetes are illustrated in Fig. 2.5 and the diversity of structures found is reflected in the recognition of nearly 1700 genera and ten times as many species. Amongst the many mechanisms for the dispersal of these propagules two major strategies are worth commenting on:

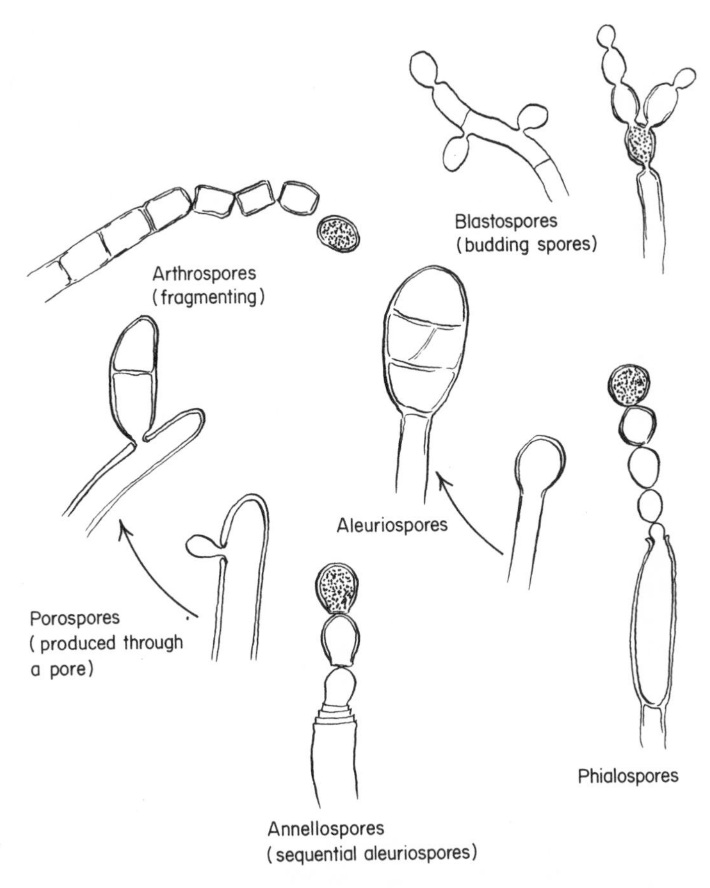

Fig. 2.5. Processes of formation of conidiospores

1. The production of small dry spores readily distributed into the atmosphere by the slightest physical disturbance. Moulds producing these spores are frequently encountered as laboratory contaminants on media exposed to the atmosphere. They include such genera as *Cladosporium*, *Penicillium*, *Paecilomyces* and *Aspergillus*.
2. The production of easily wettable spores which are not dislodged into the air without the help of rain splash or drifting droplets of water. Such spores are often thin-walled and hyaline and the species producing them are not frequently encountered as laboratory contaminants. They include the important toxigenic genus *Fusarium*.

The three genera *Aspergillus*, *Penicillium* and *Fusarium* are considered to be the most significant toxigenic moulds at the present time, but a number of other genera do include some important toxic species and these are listed in Table 2.3 and illustrated in Fig. 2.6. The structures of the associated toxins are illustrated in Fig. 2.7.

Table 2.3. Toxigenic species of Deuteromycetes other than aspergilli, penicillia and fusaria

Species	Toxins
Alternaria alternata	Tenuazonic acid
Pithomyces chartarum	Sporidesmins [a]
Trichothecium roseum	Trichothecin
Rhizoctonia leguminicola [b]	Slaframine
Stachybotrys atra	Satratoxins
Myrothecium roridum	Roridins
Phomopsis leptostromiformis	Phomopsin

[a] The name of these toxins reflects an earlier name given to the mould: *Sporidesmium bakeri.*

[b] Strictly a member of the Mycelia Sterilia.

Alternaria alternata is a common saprophyte especially on senescent plant material, foodstuffs and textiles. It is cosmopolitan in distribution and is frequently isolated from the soil. The production of branched chains of multiseptate porospores, in which each spore develops a beak-like protuberance which acts as a new conidiophore from which the next spore is produced, is characteristic of *Alternaria.*

Pithomyces chartarum is another saprophyte occurring especially on the dead leaves of fodder grasses. Although it has a worldwide distribution, it is only in New Zealand, and occasionally in Australia and South Africa, that it grows as the dominant saprophyte producing large numbers of aleuriospores each attached to a very short hyaline conidiophore. When the brown, multiseptate spores are dislodged they usually take a small fragment of the conidiophore with them and this feature is a useful aid in the identification of this genus. It is the spores which contain the toxic sporidesmins, consumption of which causes facial eczema in sheep.

Rhizoctonia leguminicola is a pathogen of red clover and does not produce conidiospores. Most, if not all, members of this genus of the Mycelia Sterilia are mycelial stages of Basidiomycetes. The production of the simple alkaloid slaframine by this mould causes a disease of cattle known as slobbers or red clover sickness.

Myrothecium roridum, *Trichothecium roseum* and *Stachybotrys atra* are all saprophytes growing on dead or damaged plant material and all three produce toxic metabolites belonging to the trichothecene group.

Phomopsis leptostromiformis produces its conidia inside a structure known as a pycnidium which has a superficial similarity to the perithecia produced by sexual reproduction in the Pyrenomycetes. The black pycnidia are produced on the stems, pods and even the seeds of infected lupin plants on which the fungus grows as both parasite and saprophyte.

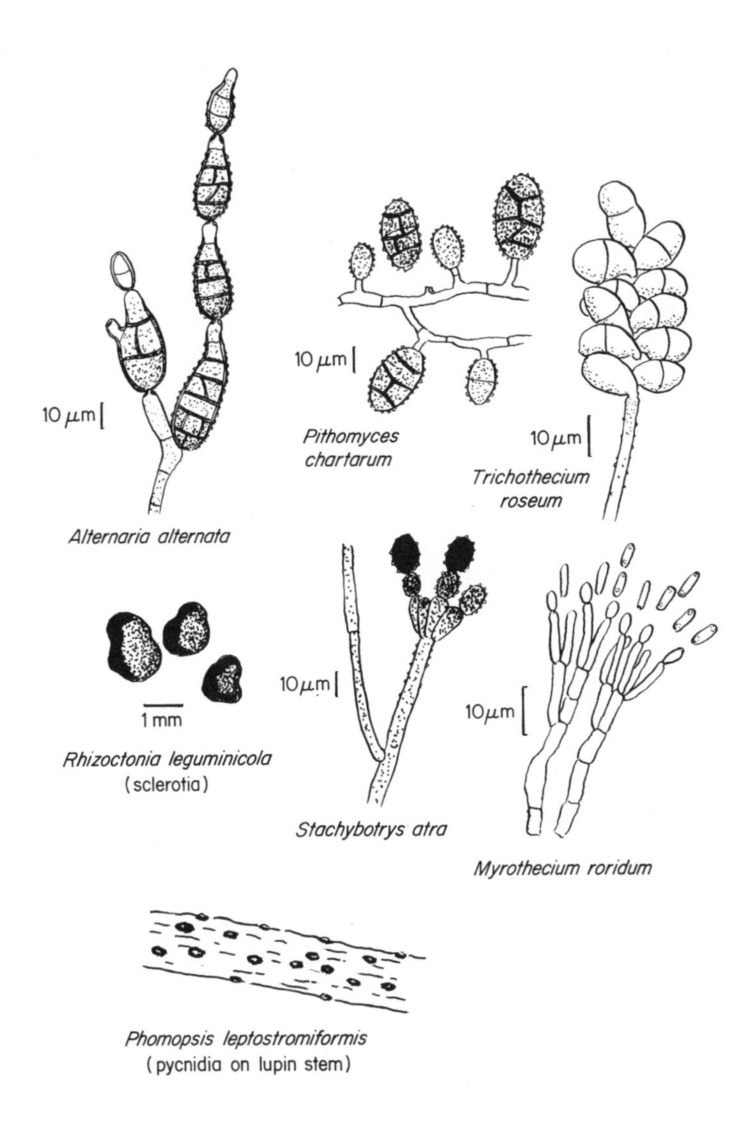

Fig. 2.6. Deuteromycetes other than penicillia, aspergilli and fusaria

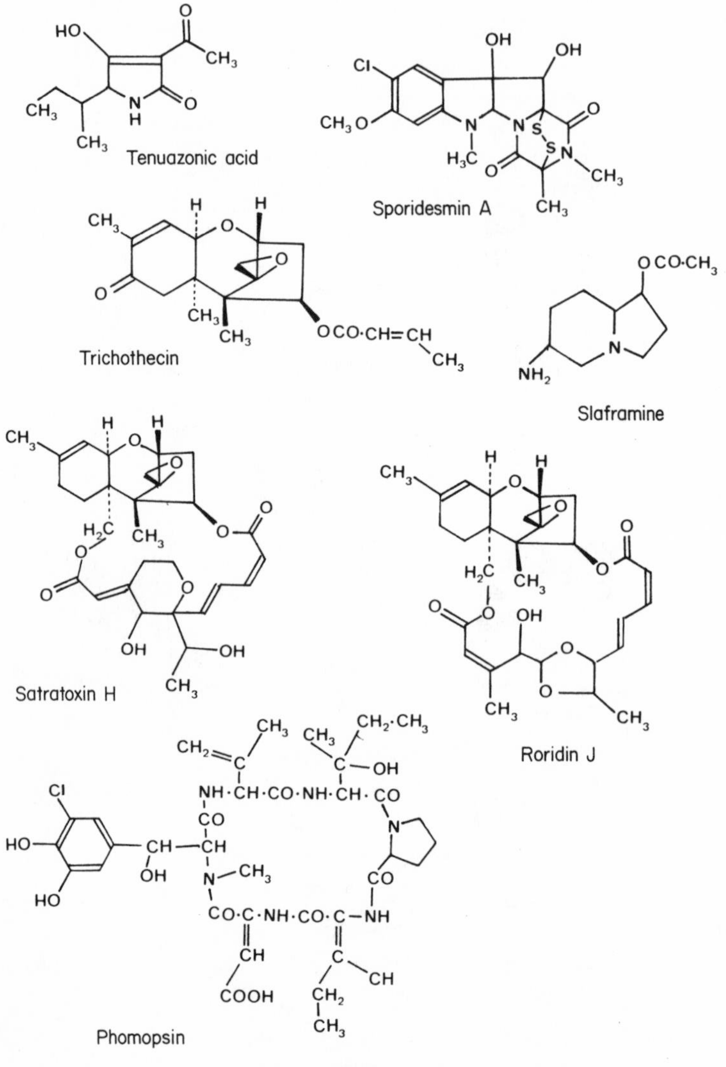

Tenuazonic acid

Sporidesmin A

Trichothecin

Slaframine

Satratoxin H

Roridin J

Phomopsin

Fig. 2.7. The structures of some toxins produced by Deuteromycetes other than penicillia, aspergilli and fusaria

The genus *Penicillium*

This genus, which includes some of the most frequently encountered moulds, is characterized by the production of small, dry, single-celled, air-dispersed conidiospores from phialides arranged as a brush-like structure on the ends of aerial conidiophores (Fig. 2.8). In temperate climates these are the dominant blue and green moulds associated with the spoilage of foods. Although essentially saprophytic, some species show some specificity in their ability to colonize certain fruits and vegetables. Thus *P. digitatum* and *P. italicum* are the green and blue moulds respectively of citrus fruits while *P. expansum* attacks apples and *P. gladioli* attacks corms and bulbs including onions.

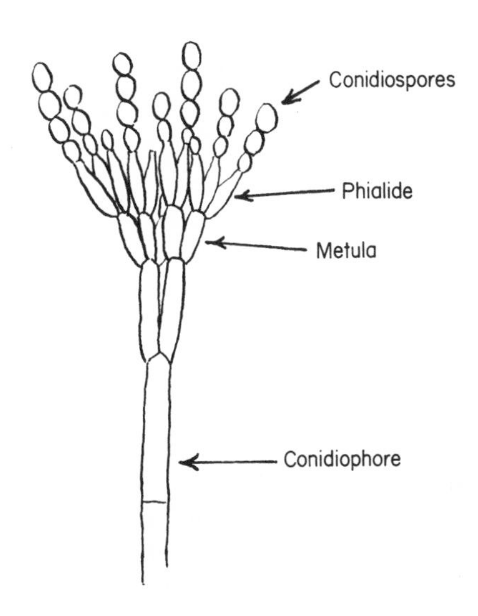

Fig. 2.8. Structure of the conidiophore of *Penicillium*

Some species of *Penicillium* are useful to man in both food and antibiotic production. *P. roquefortii* and *P. camembertii* are used in the manufacture of mould-ripened cheeses while *P. chrysogenum* is the source of much of the raw material of the penicillin industry.

The identification of individual species of *Penicillium* requires careful examination of the colour, texture and size of colonies, growing under defined conditions, as well as a detailed microscopic examination of the spore-producing structures (the penicillus). Some of the more important toxigenic species of *Penicillium* are listed in Table 2.4.

Table 2.4. The more important toxigenic species of *Penicillium* on cereals and other foods

Species	Toxins	Comments
P. citrinum	Citrinin	Common biodeteriogen, worldwide on foods, decaying plant materials, textiles
P. cyclopium	Penitrem A Cyclopiazonic acid Penicillic acid Ochratoxin A	(=*P. aurantiogriseum*). Common on cereals and other foods
P. expansum	Patulin Citrinin	Predominantly from rotting apples and pears, but also other fruits
P. islandicum	Luteoskyrin Islanditoxin Cyclochlorotine	Cereals, particularly in the tropics
P. purpurogenum	Rubratoxins	(=*P. rubrum*). Primarily a soil fungus associated with the decay of many substrates
P. roquefortii	P.R. toxin Roquefortine	Blue cheeses, also cool stored products
P. viridicatum	Ochratoxins Citrinin Viridicatin Xanthomegnin Viomellein	Worldwide, cereals and cereal products

The genus *Aspergillus*

Like *Penicillium*, the genus *Aspergillus* is characterized by the production of large numbers of small, dry, single-celled conidiospores but they are produced from phialides arranged over the surface of a swelling at the tip of the conidiophore (Fig. 2.9). Unlike *Penicillium*, in which the conidiophore is usually septate and morphologically very similar to the hyphae bearing it, the conidiophores of *Aspergillus* are often aseptate, thicker than the vegetative mycelium, and usually arise from a distinct 'foot cell'. Whereas the majority of species of *Penicillium* are most abundant in temperate climates, those of *Aspergillus* are more commonly associated with the spoilage of foods and other materials in the tropics and warmer climates. This genus is of particular importance because it contains many species capable of growth and metabolism at low water activities (Table 2.5) and are thus associated with the spoilage of food materials which are too dry to be attacked by other microorganisms.

Some species of *Aspergillus* are used extensively in the manufacture of foods and compounds used in the food industry. Thus *A. oryzae* is used in the manufacture of koji, an intermediate in the production of a number of oriental foods such as miso, and *A. niger* is used for the manufacture of citric

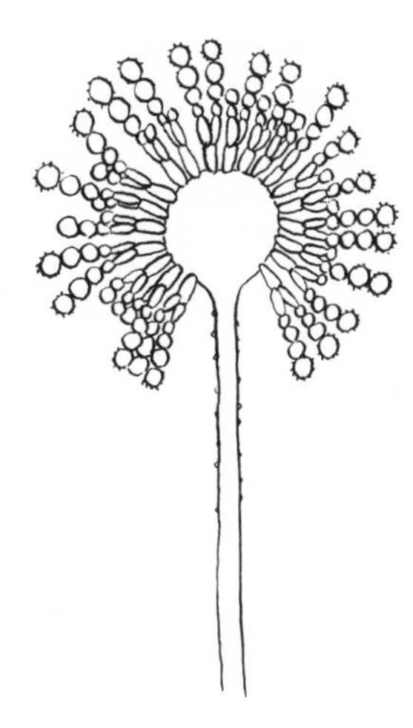

Fig. 2.9. Structure of the conidiophore of *Aspergillus*

Table 2.5. Minimum water activity (a_w) *requirements for* growth of moulds

Species	Minimum a_w [b]
Monascus bisporus [a]	0.61
Aspergillus chevalieri	0.65
Aspergillus candidus	0.72
Aspergillus versicolor	0.75
Aspergillus flavus	0.80
Aspergillus fumigatus	0.82
Aspergillus niger	0.85
Penicillium chrysogenum	0.78
Penicillium citrinum	0.80
Penicillium islandicum	0.83
Penicillium purpurogenum	0.84
Paecilomyces variotii	0.84
Fusarium spp.	0.88–0.91
Trichothecium roseum	0.90
Rhizopus and *Mucor*	0.92–0.94

[a] Previously known as *Xeromyces bisporus* and is probably the most xerophilic (xeros = dry: philos = loving) mould. It is not known to be toxigenic.

[b] $a_w = \frac{P}{P_0}$ (where P = partial presure of water vapour in equilibrium with the substrate, P_o = partial pressure of water vapour in equilibrium with pure water at the same temperature).

acid. A few species, such as *A. fumigatus*, are capable of growth in the animal body and responsible for diseases known collectively as aspergillosis, but the majority are essentially saprophytic. The genus is metabolically very versatile and includes several species producing toxic metabolites (Table 2.6).

Table 2.6. The more important toxigenic species of *Aspergillus*

Species	Toxins	Comments
A. chevalieri	Xanthocillin	Low a_w, stored cereals and cereal products
A. clavatus	Patulin Cytochalasin E Tryptoquivaline	Alkali-tolerant, animal dung, soil and decomposing organic material
A. flavus	Aflatoxins Aflatrem	Tropical and subtropical soils, plant products such as groundnuts and maize
A. fumigatus	Viriditoxin Gliotoxin Fumagillin Verruculogen	Thermophilic, decomposing organic material, pathogenic to birds and mammals
A. niger	Malformins Oxalic acid	Cosmopolitan but particularly in the tropics
A. ochraceus	Ochratoxins Penicillic acid Destruxin B	Soils, decaying vegetation, grain, adventitious pathogen
A. parasiticus	Aflatoxins	Insect pathogen, saprophyte on plant products
A. ustus	Austocystins Austamide Austdiol Brevianamide	Widespread in soil
A. versicolor	Sterigmatocystin Cyclopiazonic acid	Soil, mature cheeses, cured meats, decaying vegetation

The metabolites of *A. flavus* and *A. parasiticus*, known as the aflatoxins have been more extensively studied than any other mycotoxins. Their recognition as potent carcinogens in some animals, and possibly in man (see Chapter 5), has made them the subjects of government legislation as well as valuable tools in the study of cancer. Initially these two moulds were considered to be part of the storage flora and the problems of aflatoxin contamination as simply problems of inadequate post-harvest storage. If this were so then the occurrence of aflatoxins in food and animal feeds could be controlled by ensuring the maintenance of the correct conditions of temperature and dryness during storage and distribution. Unfortunately, the contamination of commodities such as groundnuts and maize with aflatoxin is far more complex and may involve infection of the crop in the field before it is harvested and dried. Infection with these two species of *Aspergillus* and production of aflatoxins in the field is especially associated with drought stress and insect damage, but even healthy plant tissue can become infected if fungal spores become attached to the stigma of the developing flowers. The spores may germinate and germ tubes penetrate into the developing seed tissue without causing overt damage. The groundnuts or maize kernels may then contain some aflatoxin though appearing to be healthy. Nevertheless, the highest levels of aflatoxin are formed in plant products which are visibly damaged and infected, and poorly stored, badly harvested crops are still the most likely to be contaminated with dangerously high levels of aflatoxins.

The genus *Fusarium*

Species of *Fusarium* cover a spectrum of activity from those which are fairly specific plant pathogens to those which are saprophytic on senescent plant materials or even causing biodegradation of industrial products (Thomas, 1984). Amongst the plant pathogens are species causing serious root rots, seedling deaths and canker of mature plant tissue. The genus is characterized by the production of multiseptate, hyaline macroconidia which are curved, sometimes very subtly, in the long axis (Fig. 2.10). The spores are produced from phialides and the basal cell has a more or less distinctive heel which is diagnostic for the genus.

Although there is usually little difficulty in recognizing a mould as belonging to the genus *Fusarium* if it has produced macroconidia, it is often

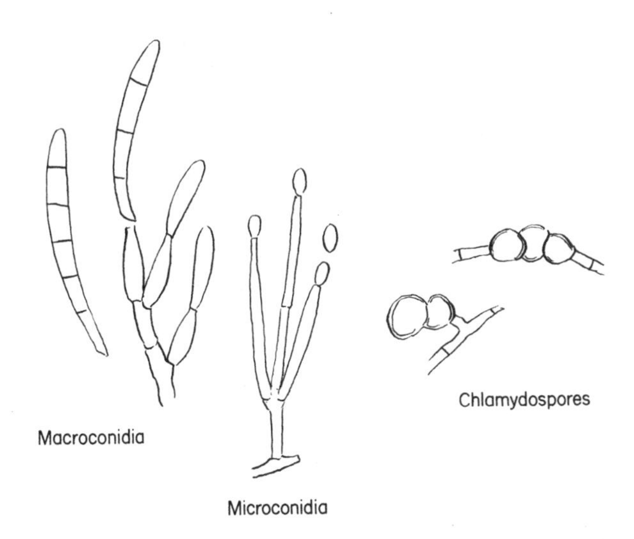

Fig. 2.10. Macroconidia, microconidia and chlamydospores of *Fusarium*

much more difficult to identify isolates to species. The production of micro-conidia and terminal or intercalary chlamydospores, the extent of growth and pigment production on a defined medium, as well as the shape and size of the macroconidia, are all characters which are used in the identification of species.

Because of their close association with plants, and their relatively high water activity requirements for growth (Table 2.5), fusaria will usually have become established on a crop before harvest and may cause a lot of problems in cereals following a late harvest after a wet summer. Although some species, such as *F. moniliforme*, are especially associated with tropical and subtropical climates and others, such as *F. sporotrichioides*, with cold climates, many species occur in temperate parts of the world.

The fusaria produce a large number of toxic sesquiterpene metabolites called trichothecenes as well as zearalenone, moniliformin, butenolide and the fusarins. Table 2.7 shows the pattern of toxin formation amongst species of *Fusarium* and the associated teleomorph, or perfect state, where this is known. As can be seen, several teleomorph genera have been associated with the anamorph genus *Fusarium* and, indeed, *Nectria* has been associated with a number of other anamorph genera which themselves do produce toxic trichothecene derivatives (Table 2.8).

Table 2.7. Toxin formation and teleomorphs of species of *Fusarium*

Species	Toxins					Teleomorph
	Trichothecenes	Zearalenone	Moniliformin	Fusarin	Butenolide	
F. moniliforme	–	+	+	+	–	*Gibberella fujikuroi*
F. oxysporum	+?	+	+	–	–	–
F. culmorum	+	+	–	–	–	–
F. avenaceum	+	+	–	–	–	*Gibberella avenacea*
F. equiseti	+	+	+	–	–	*Gibberella intricans*
F. graminearum	+	+	–	–	–	*Gibberella zeae*
F. lateritium	+	+	–	–	–	*Gibberella baccata*
F. solani	+	–	–	–	–	*Nectria haematococca*
F. nivale[a]	+	–	–	–	+	*Monographella nivalis*

[a]Also referred to as *Gerlachia nivalis.*

Table 2.8. Trichothecene-producing genera other than *Fusaria*

Anamorph genus	Toxins (some spp.)	Teleomorph (some spp.)
Myrothecium	Verrucarins, Roridin	*Nectria*
Dendrodochium	Verrucarins, Roridin	*Nectria*
Cylindrocarpon	Roridins	*Nectria*
Stachybotrys	Satratoxin, Roridin	–
Trichoderma	Trichodermin	*Hypocrea*
Trichothecium	Trichothecin	*Hypomyces*
Cephalosporium[a]	Crotocin	–
Verticimonosporium	Vertisporin	–

[a] *Cephalosporium* = *Acremonium*

ISOLATION AND CULTURE OF MOULDS

One of the problems in the isolation of moulds from infected material is that single species rarely occur alone. There will usually be a mixture of different species of moulds as well as yeasts and bacteria. Individual propagules of single-celled microorganisms or of those moulds which readily form large numbers of minute spores, may grossly outnumber the potential propagules of a significant mould growing into the tissues of a plant product. This will often mean that attempts to isolate that particular organism will be swamped by the growth of numerous colonies of other organisms. Fortunately, bacteria, being prokaryotes, are sensitive to a different range of antibiotics than are the eukaryotic fungi and there are numerous media for the selective isolation of fungi in the presence of bacteria. A widely used medium is rose-bengal agar containing tetracycline or chloramphenicol as antibacterial agents (Table 2.9). The rose-bengal restricts the growth of those moulds, such as Zygomycetes, which would otherwise quickly overgrow the whole isolation plate.

The food microbiologist is often faced with the problem of assessing the number and significance of microorganisms in a food sample. For single-celled organisms, such as bacteria and yeasts, it is possible to homogenize the sample and suspend the organisms sufficiently uniformly to allow serial dilutions to be made and plate counts obtained on the premise that each cell will produce a visible colony. Even in this situation there are problems of sampling, clumping, adhesion to particles and loss of viability during the preparation of the sample, but the problems of enumerating moulds are very much greater.

If a mould has produced spores, the technique described above will give a count reflecting the number of viable spores. This may be useful in assessing the fungal contamination of material such as ground spices. It may, however, be far more important to know something about the amount of mycelium or fungal biomass growing deep within the tissues of the substrate. If a sample of grain, such as barley or wheat, is plated out directly on an appropriate medium, it is likely that the species of fungi which grow out from the grains will be those which readily form spores which contaminate the surface, such as *Cladosporium* and *Penicillium*. If the grain is first surface sterilized with, for example, 2 per cent sodium hypochlorite solution for up to 30 mins., then washed thoroughly with sterile distilled water and plated out, those fungi which are truly growing within the tissue of the seed will grow out and can then be identified.

Table 2.9. Commonly used mycological media

Medium	Composition (per litre)	Uses
Rose-bengal agar	Glucose (10 g) Peptone (5 g) KH_2PO_4 (1 g) $MgSO_4.7H_2O$ (0.5 g) Rose-bengal (35 mg) Agar (15 g) Tetracycline (35 mg)	Initial isolation from soils, plant materials and foods
Czapek dox agar	Sucrose (30 g) $NaNO_3$ (2 g) K_2HPO_4 (1 g) $MgSO_4.7H_2O$ (0.5 g) KCl (0.5 g) $FeSO_4$ (10 mg) $CuSO_4$ (5 mg) $ZnSO_4$ (10 mg) Agar (15 g)	Identification and maintenance of *Aspergillus* and *Penicillium* (*P. digitatum* will not grow on this medium)
Malt extract agar	Malt extract (20 g) Agar (15 g)	Good general medium for mucorales and the majority of moulds
Potato sucrose agar	Potato extract (500 ml) Sucrose (20 g) Agar (15 g)	Growth and identification of *Fusarium*

Such techniques will only give a qualitative assessment of what is present. To obtain a quantitative estimate of biomass, it is necessary to either measure the mycelium, with or without staining, by microscopic observation of macerated or sectioned material or to estimate fungal mycelium by some specific chemical assay. The compounds which have been used for the chemical assay of fungal biomass in animal or plant tissue include chitin and ergosterol. The hyphal walls of the majority of terrestrial filamentous fungi contain chitin as a major constituent. This can be hydrolyzed to its monomer, N-acetyl glucosamine, which can be assayed colorimetrically. Insects also produce chitin in their exoskeleton and so any insect damaged material is likely to give a false positive result. Another growth-associated component of the fungi is the sterol ergosterol which is readily extracted and specifically analysed by such techniques as high performance liquid chromatography (HPLC) and gas liquid chromatography (GLC). It is essential that the analysis distinguishes ergosterol from plant or animal sterols. These methods have two shortcomings: different species of fungi have a different biomass:chitin (or ergosterol) ratio, and even in a single species this ratio will vary with the conditions of growth. Nevertheless, they do provide a possibility of obtaining some quantitative indication of fungal biomass present within a solid substrate.

Media used for isolation are usually not appropriate for maintenance or identification and it is frequently necessary to use yet another medium for the production of secondary metabolites such as mycotoxins. Table 2.9 lists some of the most commonly used media and their uses and Booth (1971b) lists many more. Some identification schemes, such as that of Pitt (1979) for the penicillia, use a small number of different media and growth conditions to produce sufficient characteristics for the identification of an isolate by a synoptic key.

There are a number of techniques for preserving living moulds and each is usually appropriate for a particular group. Those which produce dry, long-lived spores can be grown up on an agar slope in Universal bottles until they have sporulated and then stored under dry, cool conditions. Although refrigeration or cold room temperatures can give extended viability, there may be problems arising from condensation. Many fungi can be preserved as soil cultures by mixing a spore suspension in sterile water with sterile, seived garden loam, allowing the culture to grow for several days, and then drying and refrigerating. This technique has been valuable for wet spored genera such as *Fusarium*. Details about both preservation and maintenance of living fungi are described by Smith and Onions (1983).

When making an intensive study of a small number of isolates it is frequently convenient to maintain actively growing cultures as sources of inoculum by mass transfers to fresh medium at fairly short intervals. However, many moulds rapidly change their characteristics when frequently subcultured in this way and one of the properties which may change is the production of secondary metabolites including mycotoxins. It is a good safeguard to ensure that some long-term maintenance cultures of the original isolate have been established and to deposit the organism in a recognized culture collection such as that at the Commonwealth Mycological Institute, Kew.

FURTHER READING

Booth, C. (1971a). *The Genus Fusarium*, Commonwealth Mycological Institute, Kew.

Booth, C. (1971b). Fungal culture media. In: *Methods in Microbiology*, vol. 4, C. Booth (Ed.), Academic Press, London, pp. 49–94.

Burnett, J. H. and Trinci, A. P. J. (Eds). (1979). *Fungal Walls and Hyphal Growth*, Cambridge University Press, Cambridge.

Ellis, M. B. (1971). *Dematiaceous Hyphomycetes,* Commonwealth Mycological Institute, Kew.
Ellis, M. B. (1976). *More Dematiaceous Hyphomycetes,* Commonwealth Mycological Institute, Kew.
Gallagher, R. T., White, E. P. and Mortimer, P. H. (1981). Ryegrass staggers: isolation of potent neurotoxins lolitrem A and lolitrem B from staggers-producing pastures. *New Zealand Veterinary Journal,* **29**, 188–190.
Hawksworth, D. L., Sutton, B. C. and Ainsworth, G. C. (1983). *Ainsworth & Bisby's Dictionary of the Fungi,* 7th edn, Commonwealth Mycological Institute, Kew.
Moss, M. O. and Smith, J. E. (Eds). (1984). *The Applied Mycology of Fusarium,* Cambridge University Press, Cambridge.
Nelson, P. E., Tousson, T. A. and Marasas, W. F. O. (1983). *Fusarium Species, An Illustrated Manual for Identification,* Pennsylvania State University Press, University Park.
Onions, A. H. S., Allsopp, D. and Eggins, H. O. W. (1981). *Smith's Introduction to Industrial Mycology,* Edward Arnold, London.
Pitt, J. I. (1979). *The Genus Penicillium and its Teleomorphic States, Eupenicillium and Talaromyces,* Academic Press, London.
Raper, K. B. and Fennell, D. I. (1965). *The Genus Aspergillus,* Williams and Wilkins, Baltimore.
Raper, K. B. and Thom, C. (1948). *A Manual of the Penicillia,* Williams and Wilkins, Baltimore. (A facsimile copy of this valuable book was produced by Hafner, New York, 1968.)
Smith, D. and Onions, A. H. S. (1983). *The Preservation and Maintenance of Living Fungi,* Commonwealth Mycological Institute, Kew.
Sutton, B. C. (1980). *The Coelomycetes, Fungi Imperfecti with Pycnidia, Acervuli and Stromata,* Commonwealth Mycological Institute, Kew.
Thomas, J. L. (1984). *Fusarium* as a biodeteriogen: a case history. In: *The Applied Biology of Fusarium,* M. O. Moss and J. E. Smith (Eds), Cambridge University Press, Cambridge, pp. 107–116.
Von Arx, J. A. (1979). Propagation in the yeasts and yeast-like fungi. In: *The Whole Fungus,* B. Kendrick (Ed.) National Museum of Canada, Ottawa, pp. 555–571.
Webster, J. (1980). *Introduction to Fungi,* 2nd edn, Cambridge University Press, Cambridge.

CHAPTER 3

Structure and Formation of Mycotoxins

The chemical structures of mycotoxins have provided many challenging problems for the natural products chemist and a diverse range of complex molecules are now known. Although it is difficult to catalogue and classify these compounds by their chemical structures alone, it is possible to see relationships on the basis of biosynthetic pathways because they are all secondary metabolites.

SECONDARY METABOLISM

All forms of life, including moulds, require exogenous materials to build into biomass. As heterotrophs the moulds require organic compounds for both the synthesis of biomass (anabolic metabolism) and to produce the energy to drive these reactions (catabolic metabolism). These aspects of metabolism are frequently referred to as primary metabolism and many of the general features of such processes are shared by many groups of organisms.

Secondary metabolism is distinct from primary metabolism in so far as:

1. It occurs optimally after a phase of balanced growth (Fig. 3.1) and is often, but not always, associated with morphogenetic changes such as sporulation;
2. The production of particular secondary metabolites is usually restricted to a small number of species and may be species, or even strain, specific;
3. It has not generally been possible to rationalize the biological function of secondary metabolites, although some are very active against microorganisms (antibiotics), plants (phytotoxins) or animals (mycotoxins).

Secondary metabolites are relatively small, complex organic compounds the structures of many of which are documented in two volumes of an encyclopaedic study by Dr W. B. Turner and his colleagues of Imperial Chemical Industries (Turner, 1971; Turner and Aldridge, 1983). These books alone demonstrate the enormous diversity of chemical structures to be found but new structures are being elucidated and reported in virtually every new issue of the relevant journals.

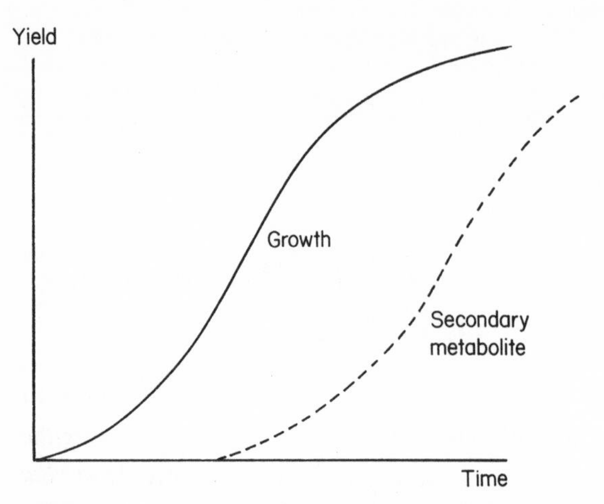

Fig.3.1. The relationship between growth and secondary metabolism

Although secondary metabolites in general, and mycotoxins specifically, do not form a neat and recognizable group of organic structures, they can be classified in terms of the biosynthetic pathways leading to their production. This is so because the processes of primary and secondary metabolism are linked by a relatively small number of simple intermediates such as acetyl coenzyme A, mevalonic acid and amino acids (Fig. 3.2). It frequently happens that a particular secondary metabolite is itself a precursor for a series of reactions leading to the formation of further groups of compounds. It is commonly observed that the more steps there are leading to the formation of a mycotoxin, the more limited is the number of species producing it. This is illustrated by the biosynthesis of aflatoxin G via versicolorin and sterigmatocystin (Table 3.1).

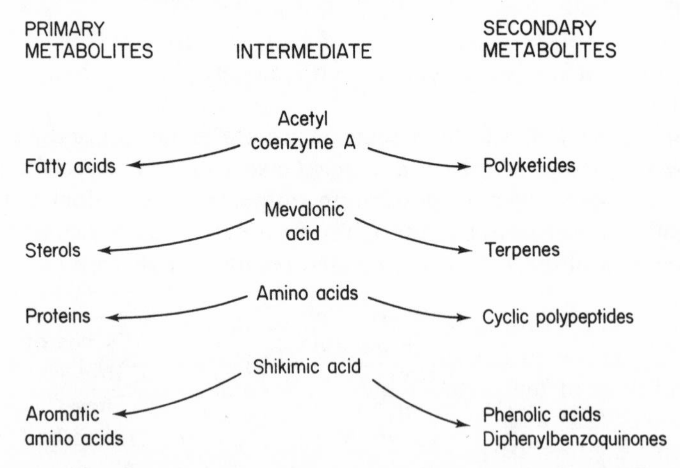

Fig. 3.2. Intermediates linking primary and secondary metabolism

Table 3.1. Range of taxa producing intermediates associated with the biosynthesis of the aflatoxins (see Fig. 3.8)

Intermediate	Taxa [a]
Averantin-type anthraquinones	*Aspergillus flavus* group
	Aspergillus nidulans group
	Aspergillus ustus group
	Aspergillus versicolor group
	Cercospora smilacis
	Dothistroma pini
	Monocillium nordinii
	Solorinia crocea
Versicolorin-type anthraquinones	*Aspergillus flavus* group
	Aspergillus nidulans group
	Aspergillus ustus group
	Aspergillus versicolor group
	Bipolaris sp.
Sterigmatocystin-type xanthones	*Aspergillus flavus* group
	Aspergillus glaucus group
	Aspergillus nidulans group
	Aspergillus ustus group
	Aspergillus versicolor group
	Bipolaris sorokimana
	Chaetomium thielavioideum
	Farrowia sp.
Aflatoxins	*Aspergillus flavus* group

[a] Many species of *Aspergillus* have been studied and, for this table, they have been reduced to the species groups recognized in Raper and Fennell (1965).

POLYKETIDE PATHWAY

During the biosynthesis of fatty acids, which are important primary metabolites used for storing chemical energy as well as forming part of the complex membranes surrounding cells and organelles, acetyl coenzyme A and a number of molecules of malonyl coenzyme A are coupled together in a regular fashion (Fig. 3.3). Although the results can be formally visualized as the linking together of acetyl groups in a head to tail fashion, the role of both carbon dioxide to form the malonyl derivative and of coenzyme A are very important in ensuring that the biosynthesis takes place under the mild conditions of temperature and pH present in the cytoplasm. For the production of fatty acids the ketone groups formed as each C2 unit is added are reduced continuously to give the paraffin chain of the final product. If these ketone groups are not reduced the resulting compound will be very reactive and can undergo a series of condensation reactions often leading to the formation of ring compounds. Such intermediates have been called polyketides and they are conveniently classified as tri-, tetra-, pentaketides, etc., depending on the number of acetyl groups incorporated. The relationship of the polyketide pathway to a number of mycotoxins is shown in Table 3.2 where it can be seen that there are representatives in all the groups from tetra- to decaketides.

Table 3.2. Polyketide-derived mycotoxins

Group	Mycotoxins	Genera
Tetraketides	Patulin	*Penicillium, Aspergillus, Paecilomyces*
	Penicillic acid	*Penicillium*
Pentaketides	Citrinin	*Penicillium*
	Ochratoxin	*Aspergillus, Penicillium*
	Austdiol	*Aspergillus*
Hexaketides	Maltoryzine	*Aspergillus*
Heptaketides	Viomellein	*Aspergillus, Penicillium*
	Viopurpurin	*Trichophyton, Penicillium*
	Rubrosulphurin	*Aspergillus, Penicillium*
	Xanthomegnin	*Trichophyton, Aspergillus, Penicillium*
	Citromycetin	*Penicillium*
	Alternariol	*Alternaria*
	Altenuene	*Alternaria*
	Altenuic acid	*Alternaria*
Octaketides	Luteoskyrin	*Penicillium*
	Rugulosin	*Penicillium*
	Rubroskyrin	*Penicillium*
	Islandicin	*Penicillium*
	Secalonic acid (= Ergochromes)	*Claviceps, Aspergillus, Phoma, Pyrenochaeta*
Nonaketides	Citreoviridin	*Penicillium, Aspergillus*
	Asteltoxin	*Aspergillus*
	Zearalenone	*Fusarium*
	Viridicatum toxin	*Penicillium*
Decaketides	Aflatoxin	*Aspergillus*
	Sterigmatocystin	*Aspergillus*
	Austocystin	*Aspergillus*

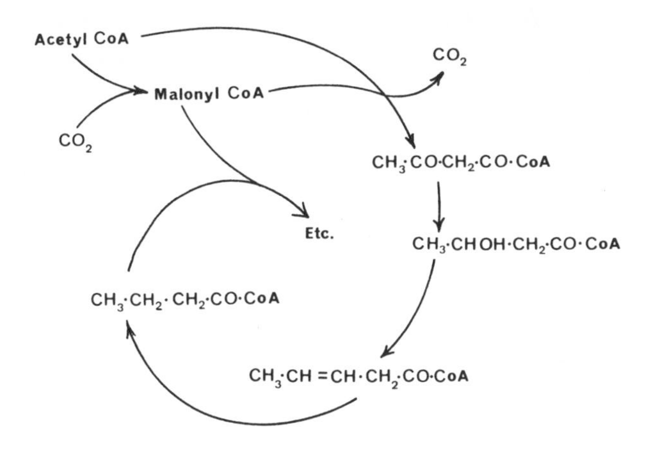

Fig. 3.3. The biosynthesis of fatty acids

The pattern of folding and condensation of the initial polyketide chain is fairly specific for any particular strain of mould and, presumably, takes place on a specific enzyme surface. It may be mediated by divalent metal ions, such as calcium, magnesium or zinc, interacting with the enol form of the polyketide chain. Whatever the final details of polyketide biosynthesis are, the initial stages are so closely related to fatty acid synthesis that it seems possible that the polyketide synthetase enzyme complex has evolved from the fatty acid synthetase complex.

The majority of polyketide-derived mycotoxins involve rearrangement of the initial condensation product. Thus, in the case of patulin, the tetraketide chain condenses to form 6-methyl salicylic acid which is then decarboxylated, oxidized, ring opened and recyclized to form the mycotoxin (Fig. 3.4). The elucidation of pathways such as this has been made possible by the use of radioactively labelled acetate using the ^{14}C isotope of carbon and, more recently, by using the non-radioactive isotope ^{13}C and nuclear magnetic resonance spectroscopy. A typical incorporation pattern is shown in Fig. 3.5 for the biosynthesis of another tetraketide-derived mycotoxin, penicillic acid.

An alternative manner in which diversification of mycotoxins may arise from a single polyketide chain is illustrated by the formation of ochratoxin, austdiol and citrinin from a pentaketide precursor. In these cases a different number of C_1 substituents are introduced into the original pentaketide skeleton (Fig. 3.6).

Although the biosynthesis of many polyketide-derived mycotoxins involves oxidative processes, the oestrogenic mycotoxin, zearalenone, produced by a number of species of *Fusarium*, is a partially reduced nonaketide with a relatively simple folding pattern (Fig. 3.7).

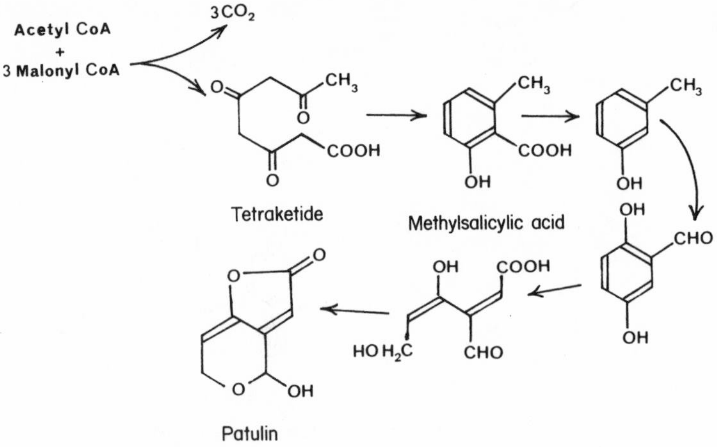

Fig. 3.4. Biosynthesis of patulin

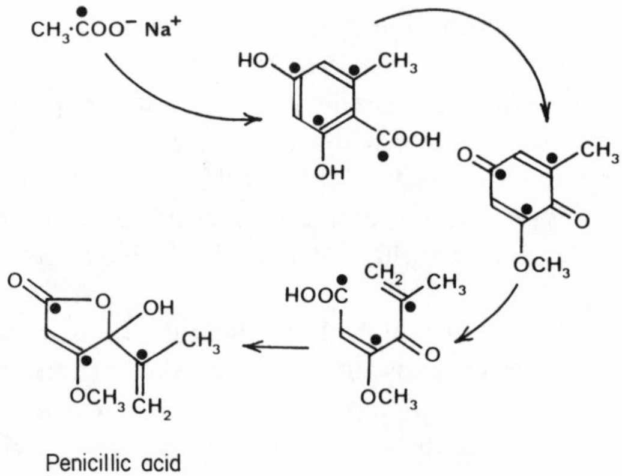

Fig. 3.5. ^{14}C distribution in penicillic acid

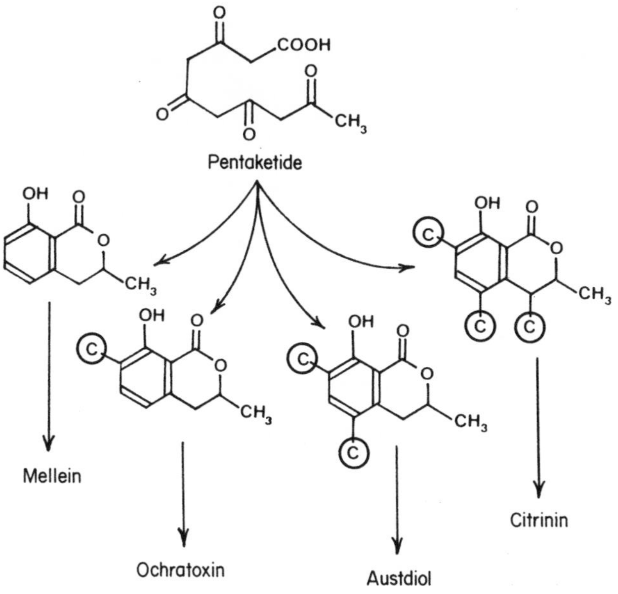

Fig. 3.6. C-methyl group incorporation into a pentaketide in the biosynthesis of citrinin, austdiol and ochratoxin

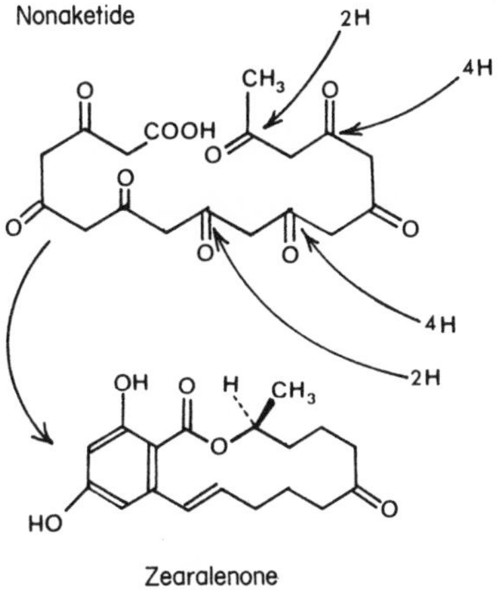

Fig. 3.7. Biosynthesis of zearalenone

The biosynthetic route to the aflatoxins has presented a particularly challenging problem to chemists, biochemists and fungal geneticists. Although the structures suggest an altered nonaketide, the derivation of the unusual bis-furan ring system was difficult to unravel. In the event, the detailed studies of many scientists throughout the world have demonstrated that the aflatoxins are derived from a decaketide from which a C_2 unit is lost during the formation of the bis-furan rings (Fig. 3.8). The sequence of intermediates included in this pathway has been elucidated with the use of mutant strains blocked at various stages, the enzyme inhibitor dichlorvos and the incorporation of radioactively labelled precursors.

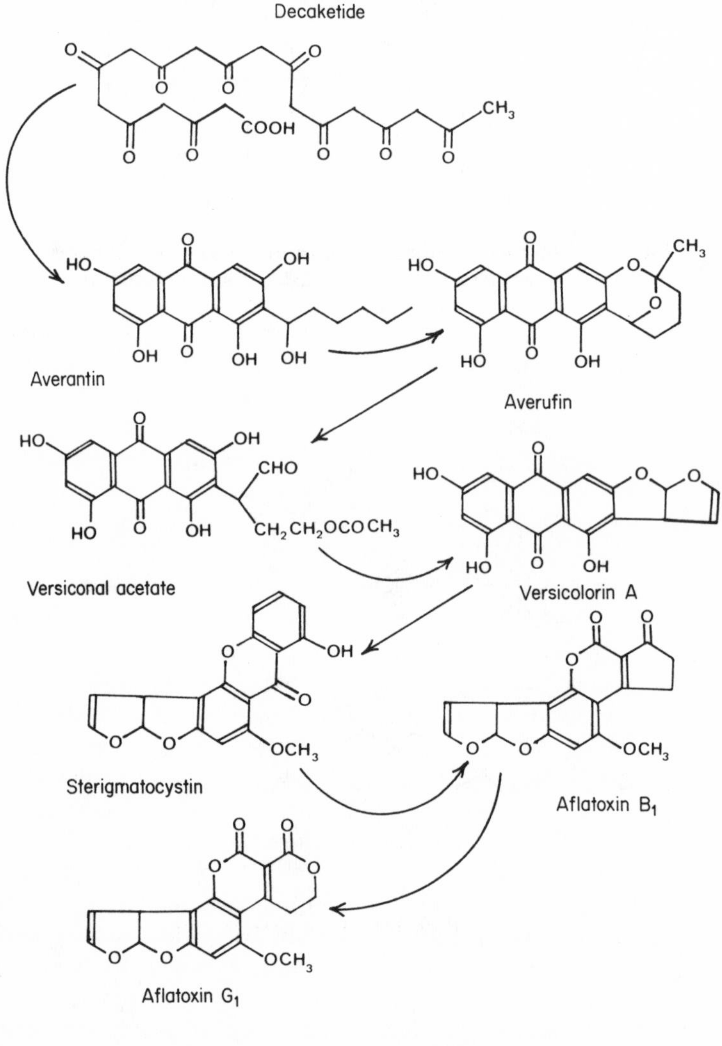

Fig. 3.8. Biosynthesis of aflatoxins

MEVALONATE PATHWAY

Mevalonic acid is an important intermediate derived from three molecules of acetyl coenzyme A (Fig. 3.9). It undergoes pyrophosphorylation, decarboxylation and dehydration to form the active isopentenyl pyrophosphate precursor of such key compounds as sterols, steroids and carotenoids. This same intermediate is also the starting point for the biosynthesis of a group of mycotoxins collectively referred to as trichothecenes. These secondary

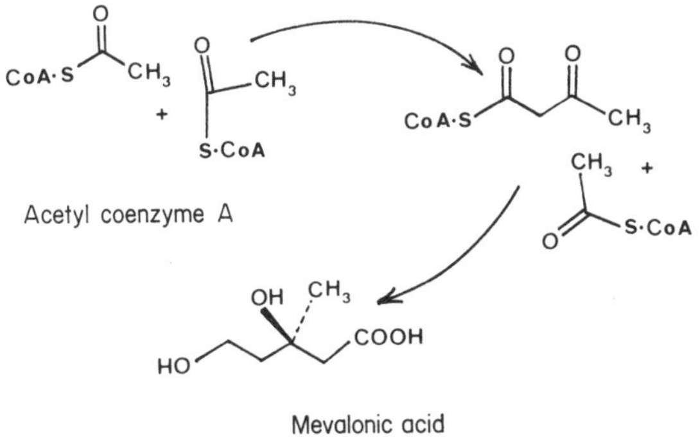

Fig. 3.9. Biosynthesis of mevalonate

metabolites belong to a family of compounds known as terpenes which are conveniently divided into monoterpenes, sesqui-, di- and triterpenes based on C_{10} C_{15} C_{20} and C_{30} skeletons, respectively. The trichothecenes are a large family of toxins all having in common a complex sesquiterpene nucleus derived from farnesyl pyrophosphate by a series of cyclizations and methyl group migrations (Fig. 3.10).

Although some members of this family of biologically active compounds had been isolated as long ago as 1946, it was not until 1964 that the correct structure was established, by X-ray crystallography of the *p*-bromobenzoate of trichodermol, the parent alcohol of trichodermin isolated from *Trichoderma viride*. Since then, many derivatives of the unusual 12,13 epoxy-trichothec-9-ene nucleus have been isolated (Table 3.3), especially from the genus *Fusarium*.

The epoxy trichothecene nucleus has five positions which may be substituted with oxygen functions (Fig. 3.11) and each of them may, or may not, be acylated. Members of the genus *Fusarium* are known to be able to hydroxylate such compounds as steroids and are certainly able to hydroxylate the trichothecene nucleus. If we simply consider the fully substituted pentahydric alcohol, it is possible to obtain 31 distinct acetate derivatives (1-penta-, 5-tetra-, 10-tri-, 10-di- and 5-monoacetates). It should, thus, be no surprise to learn that a large number of trichothecene derivatives have been isolated and characterized. The structures of a representative selection of the most studied trichothecenes are shown in Fig. 3.12.

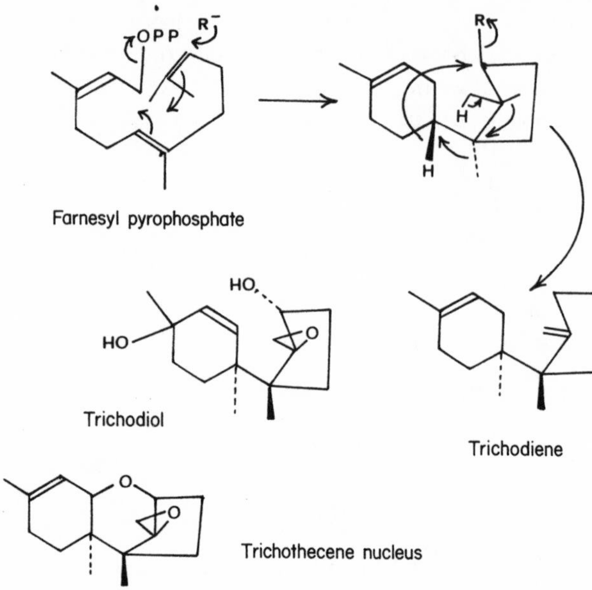

Farnesyl pyrophosphate

Trichodiol

Trichodiene

Trichothecene nucleus

Fig. 3.10. Biosynthesis of the trichothecene nucleus

Table 3.3. A selection of trichothecenes and their sources

Trichothecin	Genera
Trichodermol	*Myrothecium*
Trichodermin	*Trichoderma*
Trichothecin	*Trichothecium*
Crotocin	*Cephalosporium*
Diacetoxyscirpenol	
T-2 Toxin	
Nivalenol	*Fusarium*
Deoxynivalenol (= vomitoxin)	
Fusarenon	

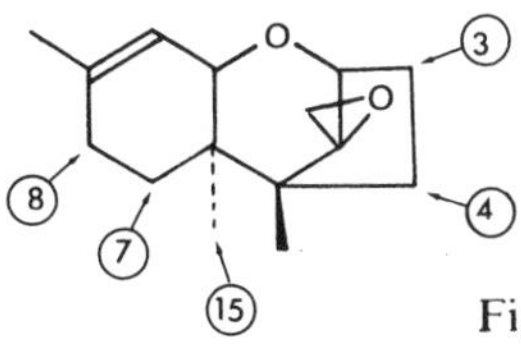

Fig. 3.11. Substitution pattern in the trichothecene nucleus

Diacetoxyscirpenol (DAS)

T-2 toxin

Deoxynivalenol (vomitoxin)

Verrucarin A

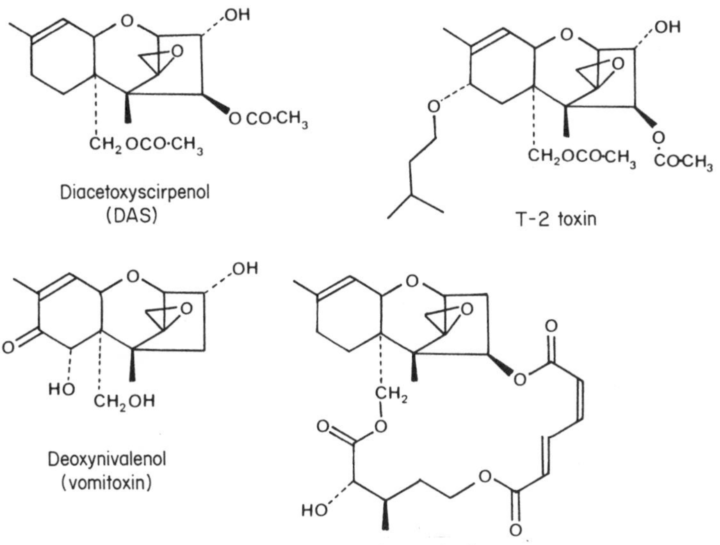

Fig. 3.12. Structures of widely studied trichothecenes

The macrocyclic compounds, in which two of the hydroxyl groups on the trichothecene nucleus are linked by a long di- or tri-ester chain to form a large ring (e.g. verrucarin A, Fig. 3.12), are not produced by any members of the genus *Fusarium* but by species from a number of other genera (Table 3.4). It has been considered that a group of such compounds, known as the baccharinoids, may be produced by the South American plant *Baccharis megapotamica*, but it is now recognized that these compounds may actually be derived by the plant metabolizing fungal macrocylic trichothecenes produced in the soil and absorbed by the roots. This would not be the only example of such a relationship between soil fungi and higher plants, although the phenomenon has not previously been observed with the trichothecenes. Neither is it unknown for moulds and higher plants to be able to synthesize the same metabolites. The sesquiterpene, abscisic acid, is an important plant hormone which is also produced by the mould *Cercospora rosicola*. Probably the best known family of diterpenes, the gibberellins, also contains many compounds which are produced by higher plants and by strains of *Fusarium moniliforme*. The structures of these two interesting metabolites are shown in Fig. 3.13.

Table 3.4. A selection of macrocyclic trichothecins and their sources

Trichothecin	Genera
Verrucarins	*Myrothecium*
Roridins	*Myrothecium*
Satratoxins	*Strachybotrys*
Iso roridin E	*Cylindrocarpon*
Vertisporin	*Verticimonosporium*

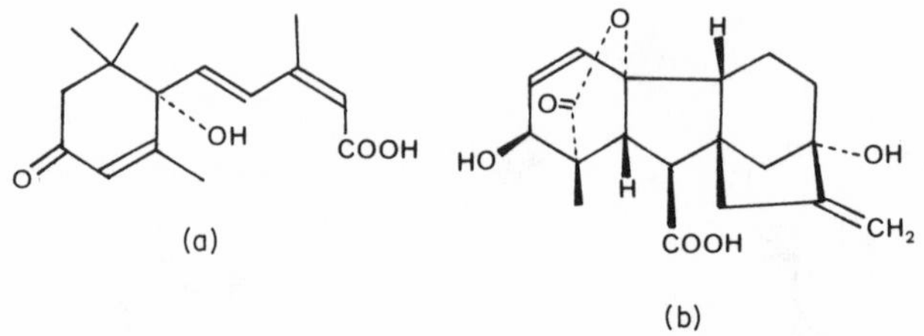

Fig. 3.13. (a) Abscissic acid and (b) gibberellic acid

A further example of a sesquiterpene-derived mycotoxin is PR-toxin produced by *Penicillium roquefortii.* An outline of its biosynthesis from farnesyl pyrophosphate, itself derived from three molecules of mevalonate, is shown in Fig. 3.14.

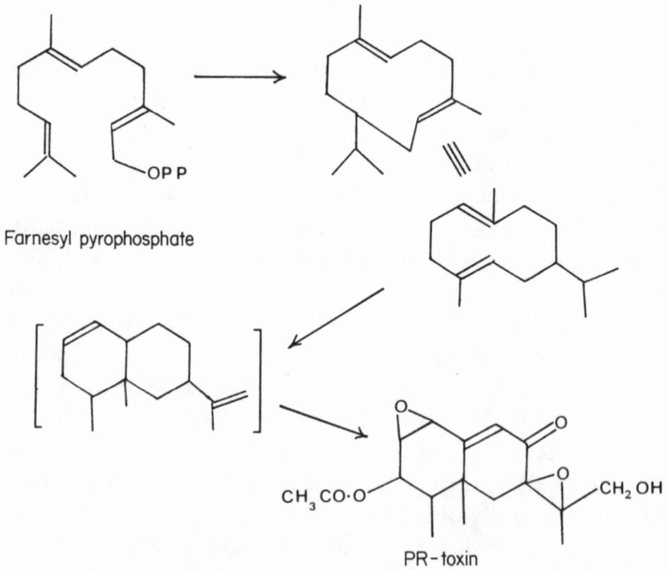

Fig. 3.14. Biosynthesis of PR-toxin

CYCLIC PEPTIDES AND THEIR DERIVATIVES

A number of mycotoxins are produced by the incorporation of amino acids into either macrocyclic peptides such as malformin C and islanditoxin (= cyclochlorotine) (Fig. 3.15) or highly condensed polycyclic compounds such as sporidesmin, gliotoxin and ergotamine (Fig. 3.16). Gliotoxin and sporidesmin share an unusual feature, found in several other mould metabolites, in the presence of a disulphide bridge, the biosynthetic origin of which is still uncertain. Whereas gliotoxin is formed from phenylalanine and serine, sporidesmin is produced from tryptophan and alanine (Fig. 3.17) and the formation of both may proceed via an intermediate containing an epoxide ring in its structure.

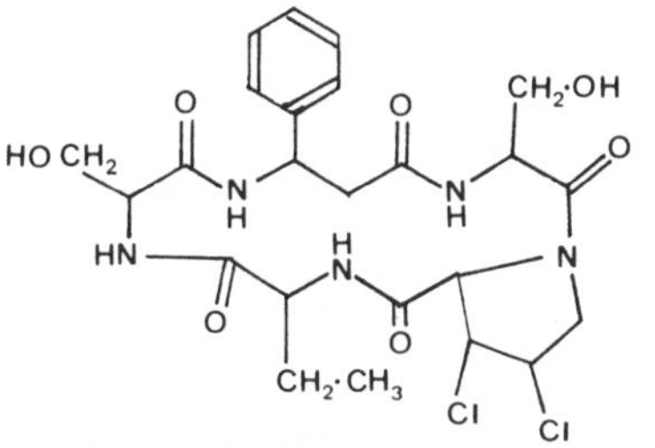

Cyclochlorotine
(= islanditoxin)

Fig. 3.15. Structures of macrocyclic polypeptides

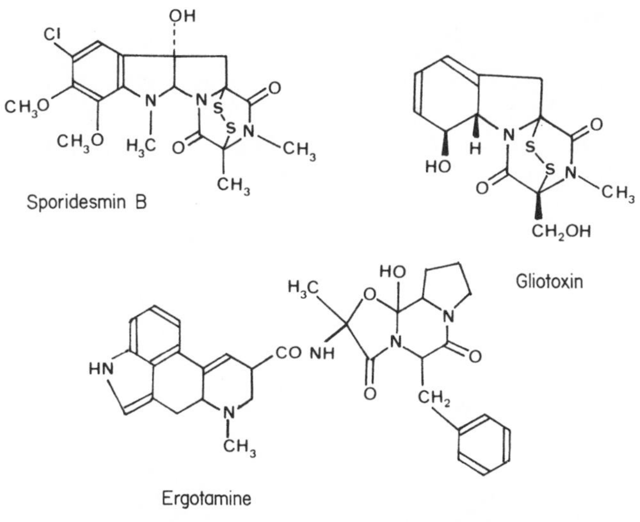

Sporidesmin B

Gliotoxin

Ergotamine

Fig. 3.16. Structures of polycyclic peptide derivatives

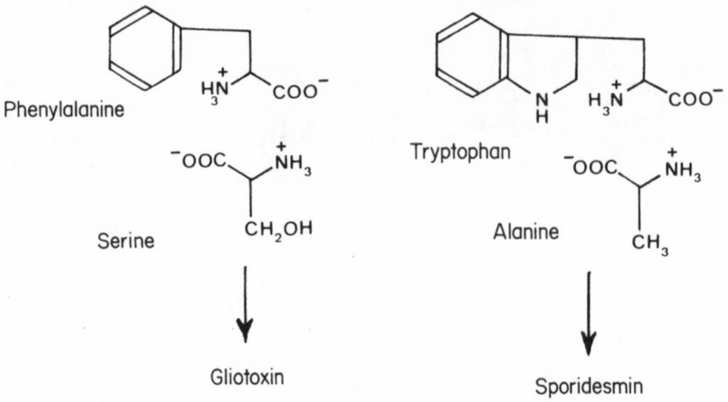

Fig. 3.17. Biosynthesis of gliotoxin and sporidesmin

The macrocyclic peptides are not produced by the machinery for normal protein biosynthesis, i.e. with the involvement of ribosomes and messenger RNA. These compounds frequently include unusual amino acids, such as D-isomers in malformin C and β phenylalanine in cyclochlorotine, which are not incorporated into proteins.

MYCOTOXINS DERIVED FROM AMINO ACIDS AND MEVALONATE

As is so often the case in nature, man's attempts to neatly classify phenomena are never entirely satisfactory and a number of mould metabolites are best considered as derived from a mixture of two or more pools of biosynthetic precursors.

The elucidation of the structures of the tremorgenic mycotoxins has uncovered a family of compounds of considerable complexity based, essentially, on tryptophan, and, possibly, other amino acids, linked to a number of mevalonate-derived isoprene units. These isoprene units may be simply peripheral substituents or they may be involved in the production of a polycyclic skeleton. Aflatrem, a metabolite of *Aspergillus flavus*, shows examples of both situations (Fig. 3.18). Roquefortine, another toxic metabolite of *Penicillium roquefortii*, is a relatively straightforward diketopiperazine, derived from tryptophan and histidine, substituted in the indole ring with an isoprene unit (Fig. 3.19).

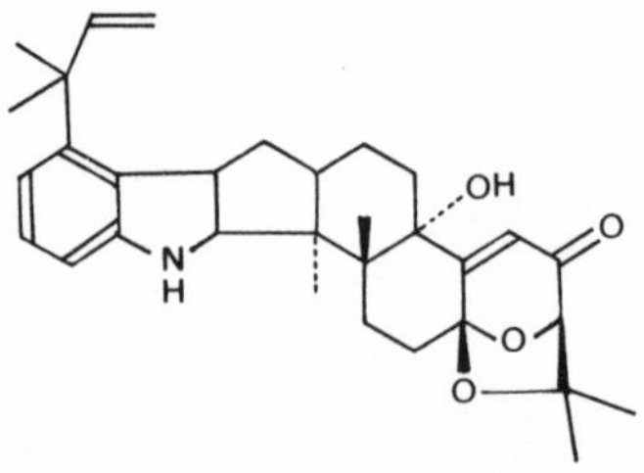

Fig. 3.18. Structure of aflatrem

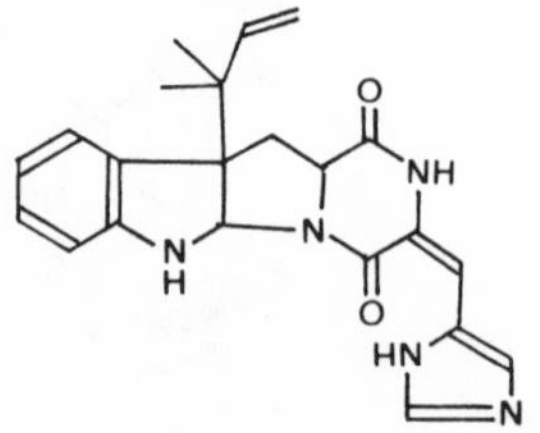

Fig. 3.19. Structure of roquefortine C

PROCESSES RELATED TO KREBS' CYCLE REACTIONS

The rubratoxins, produced by *Penicillium purpurogenum*, belong to a small group of fungal metabolites referred to collectively as nonadrides. This name reflects the presence of anhydride rings directly attached to a nine-membered carbocyclic ring (Fig. 3.20). These compounds are derived from the cyclization of two molecules of a precursor which itself is probably formed from the condensation of an activated acyl derivative and oxaloacetate (Fig. 3.21) in a reaction analogous to the citric acid synthetase reaction of the Krebs' cycle.

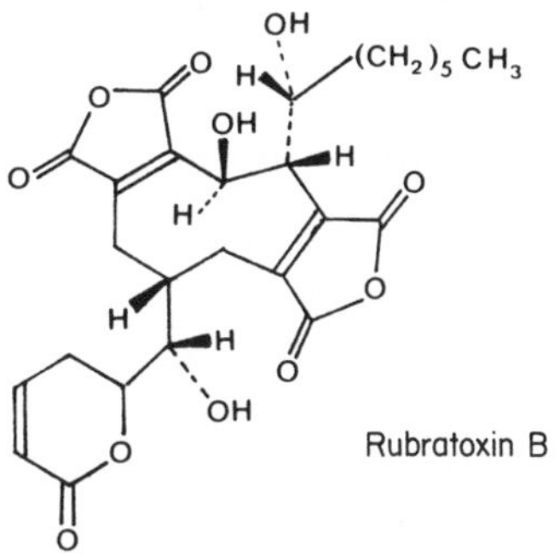

Rubratoxin B

Fig. 3.20. The nonadrides

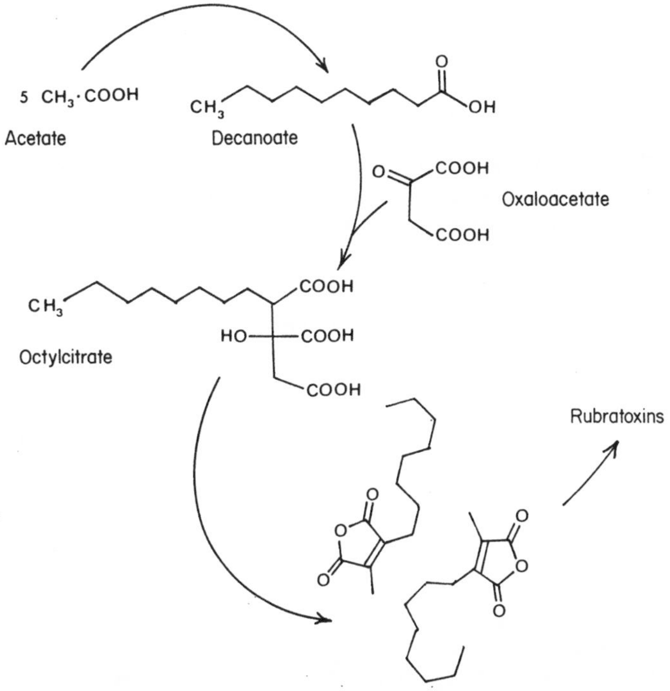

5 $CH_3 \cdot COOH$
Acetate
Decanoate
Oxaloacetate
Octylcitrate
Rubratoxins

Fig. 3.21. Postulated biosynthesis of nonadrides

THE PHYSIOLOGY OF MYCOTOXIN PRODUCTION

As already emphasized, mycotoxins are a group of fungal secondary metabolites and their production is influenced by both the genotype of the organism and the physicochemical environment in which it is growing. In view of the widespread occurrence of some species, and the usefulness of many in the manufacture of foods and food additives, it is important to appreciate that the production of any particular mycotoxin depends on the strain and not only on the species. Thus, although aflatoxins are only known to be produced by *Aspergillus flavus* and *Aspergillus parasiticus*, there are strains of both species which are non-aflatoxigenic. It would seem that *A. parasiticus* is more common on oilseeds, such as groundnuts, and frequently produces aflatoxins B_1, B_2, G_1 and G_2, whereas *A. flavus* is more common on cereals, such as maize, and usually only produces aflatoxins B_1 and B_2. Strains of both species may also produce aflatoxin M_1, the same compound that is formed in milk by cows metabolizing ingested aflatoxin B_1. Indeed, the chemical structure of aflatoxin M_1 was first elucidated using material isolated from mould growth and it was subsequently confirmed that this was identical with the material isolated from milk.

Even if a strain of mould has the genetic potential to produce a particular mycotoxin, the level of production will be influenced by the nutrients available. Very little aflatoxin is produced during the phase of vigorous growth in a laboratory culture and it is only when some nutritional factor runs out and limits growth that toxin biosynthesis occurs rapidly. The presence of amino acids, such as glutamic and aspartic acids, enhances aflatoxin biosynthesis. The presence of zinc is essential for maximum toxin production. In this instance it is possible that zinc ions stimulate glycolysis during the stationary phase, ensuring that there is sufficient acetyl coenzyme A available for polyketide biosynthesis (and, hence, aflatoxin formation) despite the fact that it is no longer required for growth.

Even when the nutritional requirements are suitable for mycotoxin biosynthesis, physical parameters, such as temperature and water activity, will influence production. In the case of *Aspergillus parasiticus*, which is essentially a subtropical species, the optimum temperature for growth is 35° C but it has been reported that maximum aflatoxin biosynthesis occurs at 25–30° C on both synthetic and natural media in laboratory experiments (Fig. 3.22). At the Coastal Plain Experimental Station, University of Georgia, USA, a sophisticated facility is available for carefully controlling soil temperature and water potential in plots on which crop plants can be grown. This facility has demonstrated that for aflatoxin production in groundnuts growing under drought conditions, the soil temperature is critical. No toxin was formed at 31.7° C, maximum levels at 29.9° C, reduced levels at 27.7° C and relatively little at 24.7° C.

It is a general observation that temperature and water activity strongly interact and this is illustrated for data recently reported for patulin biosynthesis by *Byssochlamys nivea* in apple juice (Table 3.5).

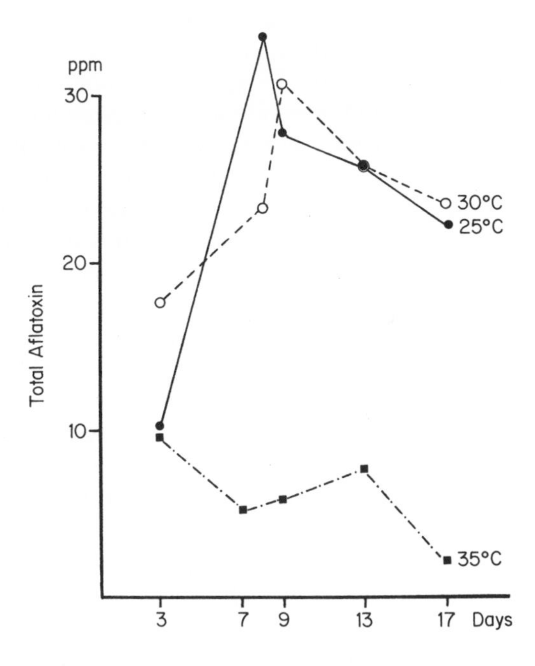

Fig. 3.22. Influence of temperature on aflatoxin production

Table 3.5. Influence of temperature and a_w on growth and patulin production by *Byssochlamys nivea* . (Data extracted from Roland and Beuchat, 1984).

	Minimum a_w at		
	21° C	30° C	37° C
Growth	0.915	0.886	0.871
Patulin	0.978	0.968	0.959

In nature there are many factors interacting with the growth and metabolism of a mould other than those indicated above. There may be, for example, antimicrobial agents produced by other microorganisms, by the plant on which the mould is growing, or added as biocides during crop husbandry. Although aflatoxin production is inhibited by some compounds, such as citric and lactic acids, which may have been produced by *Aspergillus niger* and the lactic acid bacteria, respectively, and is also inhibited by the constituents of some spices, and some biocides, it is enhanced by at least a few antifungal agents. Thus, propionic acid used to protect high moisture barley for animal feeding, and the mould metabolites rubratoxin B (from *Penicillium purpurogenum*) and cerulenin (from *Cephalosporium caerulens* and *Acrocylindrium oryzae*) all enhance aflatoxin biosynthesis even though they repress growth.

Even the plant genome is known to influence the amount of mycotoxin formed by a fungus which has successfully colonized the crop pre- or post-harvest. Indeed, one possible approach to the control of aflatoxin formation

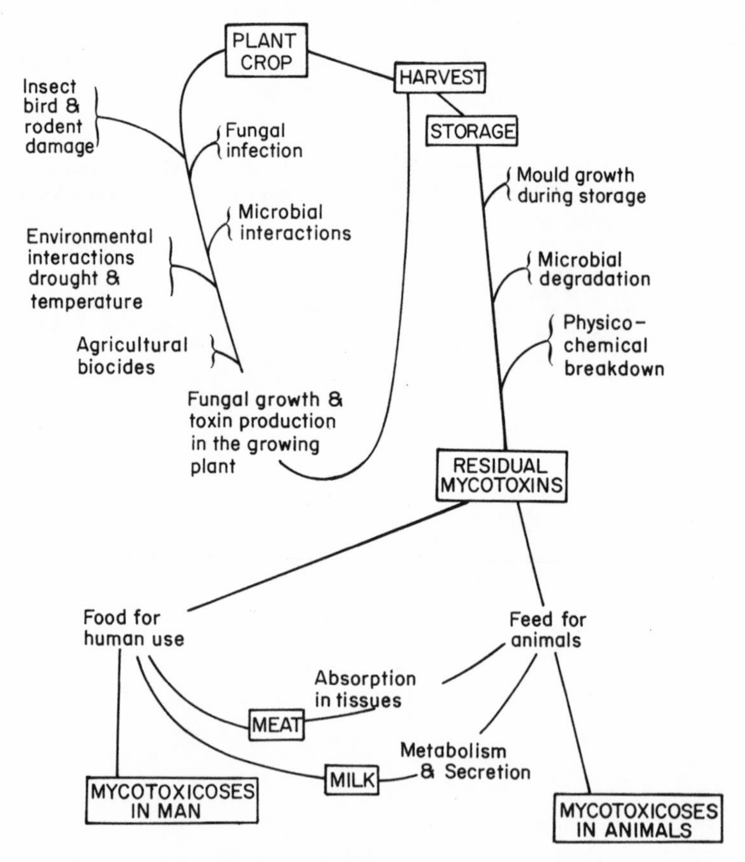

Fig. 3.23. Factors influencing the occurrence of mycotoxins in human food or animal feed

in such important crops as groundnuts and maize has been through plant breeding programmes (Mixon, 1977; Zuber, 1977). The interactions between the many possible factors that influence the occurrence of a mycotoxin in a human food or animal feed are illustrated in Fig. 3.23.

FURTHER READING

Bennett, J. W. and Ciegler, A. (Eds). (1983). *Secondary Metabolism and Differentiation in Fungi,* Marcel Dekker, New York.

Diener, U. L., Asquith, R. L. and Dickens, J. W. (Eds). (1983). *Aflatoxin and Aspergillus flavus in Corn,* Department of Research Information, Alabama Agricultural Experimental Station, Auburn University, Alabama.

Diener, U. L. and Davis, N. D. (1966). Aflatoxin production by isolates of *Aspergillus flavus. Phytopathology* **56,** 1390–1393.

Manitto, P. (1981). *Biosynthesis of Natural Products,* Ellis Horwood Publishers, Chichester.

Mixon, A. C. (1977). Influence of plant genetics on colonisation by Aspergillus flavus and toxin production (peanuts). In: *Mycotoxins in Human and Animal Health,* J. V. Rodricks, C. W. Hesseltine and M. A. Mehlman (Eds), Pathotox, Illinois.

Moss, M. O. (1984). Conditions and factors influencing mycotoxin formation in the field and during the storage of food. *Chemistry and Industry,* 533–536.

Raper, B. K. and Fennell, D. I. (1965). *The Genus Aspergillus,* Williams and Wilkins, Baltimore.

Roland, J. O. and Beuchat, L. R. (1984). Influence of temperature and water activity on growth and patulin production by *Byssochlamys nivea* in apple juice. *Applied and Environmental Microbiology* **47,** 205–207.

Steyn, P. S. (Ed.) (1980). *The Biosynthesis of Mycotoxins, a Study in Secondary Metabolism,* Academic Press, New York.

Turner, W. B. (1971). *Fungal Metabolites,* Academic Press, London.

Turner, W. B. and Aldridge, D. C. (1983). *Fungal Metabolites II,* Academic Press, London.

Zuber, M. S. (1977). Influence of plant genetics on toxin production in corn. In: *Mycotoxins in Human and Animal Health,* J. V. Rodricks, C. W. Hesseltine and M. A. Mehlman (Eds), Pathotox, Illinois.

CHAPTER 4

Implications of Mycotoxins in Animal Disease

Mycotoxins can profoundly influence the health of most animal species but their effects are particularly noticeable in assembled groups of farm animals such as dairy and feedlot cattle, pigs and poultry since their normal feeding practices involve a high intake of concentrated feeds (see Chapter 6). However, mycotoxicoses represent a diagnostically difficult problem to the veterinarian since generally the mycotoxin-induced disease syndromes are slight and can be easily confused with other diseases caused by pathogenic microorganisms or by nutrient deficiencies or imbalances. Individual mycotoxins may also affect more than one system of a diseased animal species.

Animals can demonstrate variable susceptibilities to mycotoxins depending on genetic factors (species, breed and strain), physiological factors (age, nutrition, other diseases) and environmental factors (climatic conditions, husbandry and management, etc.) (Fig. 4.1).

The effect of mycotoxins on animals has been categorized broadly into three identifiable forms (Pier *et al.*, 1980):

1. Acute primary mycotoxicoses,
2. Chronic primary mycotoxicoses,
3. Secondary mycotoxin diseases.

ACUTE PRIMARY MYCOTOXICOSES

Acute mycotoxicoses generally cause marked signs of disease or death of affected animals, with varying symptoms depending on the nature and concentration of the toxin. These effects are produced when high to moderate concentrations of mycotoxins are consumed causing a specific observable, acute disease syndrome such as hepatitis, haemorrhage, nephritis, necrosis of oral and enteric epithelia or death.

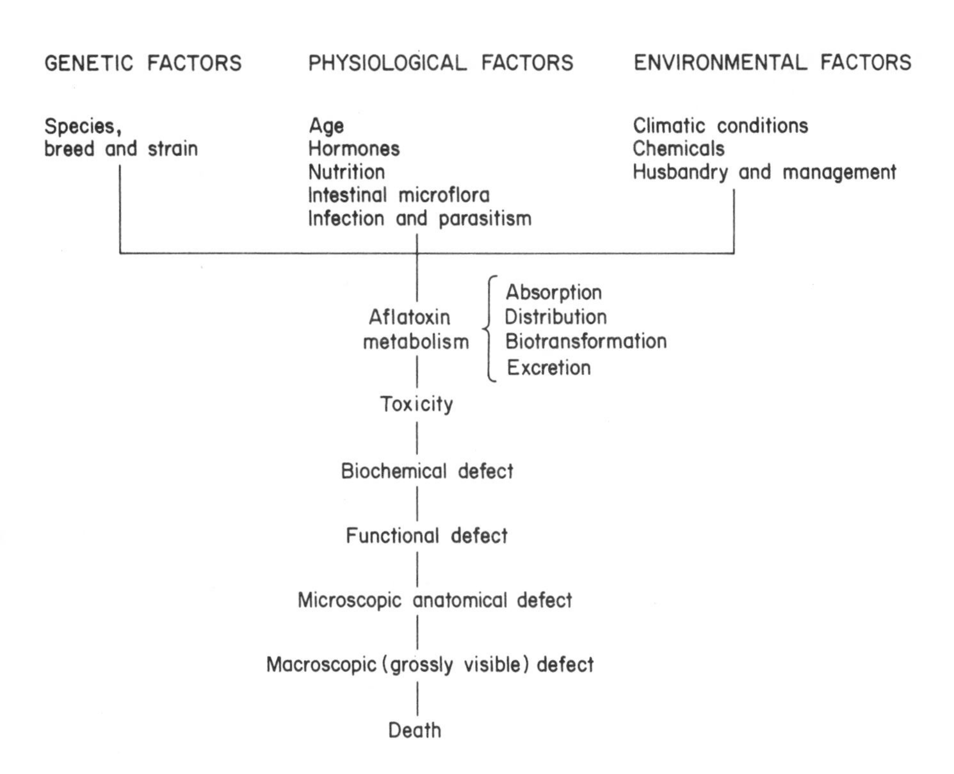

Fig. 4.1. Simplified representation of some general relationships in aflatoxicosis (Bryden, 1982)

A substance can be considered to be toxic to an animal when a single high dose or a series of small doses leads to death. The quantitative expression of toxicity, the LD_{50}, is the dosage expressed per unit body weight necessary to kill 50 per cent of a statistically valid group of animals. Levels of toxin below the LD_{50} value will produce variable degrees of sickness and chronic malfunction of the animal.

The LD_{50} values for aflatoxin B_1 and some trichothecene mycotoxins are given in Tables 1.2 and 4.1, respectively. LD_{50} values for ochratoxin A vary from 0.5–54 mg kg^{-1} body weight depending on the species of animal, for sterigmatocystin, 60–166 mg kg^{-1}, and for rubratoxin B, 60–200 mg kg^{-1}. Zearalenone, an oestrogenic mycotoxin has a low order of acute toxicity with an LD_{50} of approximately 1, 2 and 20 g kg^{-1} in mice, rats and dogs, respectively.

LD_{50} values are strongly influenced by sex, age and strain of animal, administration routes (i.e. oral feeding or intubation, intravenous or interperitoneal injection), solvents of toxins, presence of other mycotoxins, and composition of diet, etc. The data obtained will be relative and not conclusive for the evaluation of the toxicological features of individual mycotoxins.

Table 4.1. LD_{50} values (mg kg^{-1}) of selected trichothecenes (Ueno, 1983)

Type	Trichothecenes	Mouse			Rat				Chick		
									1-day old		7-day old
		i.v.	i.p.	s.c.	i.v.	i.p.	s.c.	p.o.	p.o.		
A	T-2 toxin		5.2					5.2	1.75	4.97	4.0
	HT-2 toxin		9.2						6.25	7.22	
	Diacetoxyscirpenol	12	23.0		1.3	0.75		7.3		3.82	5.0
	Neosolaniol		14.5							24.87	
B	Fusarenon-X	3.4	3.4	4.2			0.5	4.4		23.79	
	Deoxynivalenol		70.0								

Most LD_{50} data have been acquired by laboratory type experimentation. The biochemical and biological bases of the LD_{50} results are discussed later.

Pathological studies have shown that at acute levels of mycotoxins virtually every system of an animal's body can be affected by one or a combination of mycotoxins, viz:

1. Vascular system (increased vascular fragility, haemorrhage into body tissues, e.g. aflatoxin, dicoumarin).
2. Digestive system (diarrhoea, intestinal haemorrhage and hepato-toxic effects causing liver necrosis, bile duct proliferation and fibrosis, e.g. aflatoxin, caustic effects on mucous membranes; e.g. T-2 toxin, bile duct occlusion, e.g. sporidesmin, feed rejection, e.g. vomitoxin).
3. Respiratory system (adenomatosis, e.g. 4-ipomenol).
4. Nervous system (tremors, incoordination, mania, coma, e.g. tremorgens, ergotamine and related alkaloids).
5. Cutaneous system (photosensitization, e.g. sporidesmin, necrosis and sloughing of the extremeties, e.g. ergot).
6. Urinary system (nephrosis, uremia, e.g. ochratoxin, citrinin).
7. Reproductive system (infertility, prolonged oestrus, e.g. zearalenone, T-2 toxin) (Morehouse, 1979) (Tables 4.2, 4.3, 4.4 and 4.5).

All of the above listed pathological effects of mycotoxins have been observed in nature as well as with experimentally applied mycotoxins.

Normally, natural contamination levels of mycotoxins are usually not high enough to cause overt mycotoxicoses. Reports of naturally occurring outbreaks of acute mycotoxicoses are limited in both number and description. It is now becoming increasingly recognized that most field or feed contamination with mycotoxins will cause chronic mycotoxicosis symptoms or secondary mycotoxin diseases.

Table 4.2. Mycotoxicoses associated with acute primary diseases of livestock and poultry (Pier *et al.*, 1980)

Mycotoxicosis	Animal species	Primary syndrome
Aflatoxicosis	Poultry, swine, cattle, dog	Acute hepatitis; haemorrhagic disease; death
Ergotism	Cattle, sheep, chicken	Gangrenous necrosis; nervous seizures; reproductive failure
Facial eczema	Sheep, cattle	Cholangiohepatitis; photosensitivity
Fusariotoxicoses		
Vomitoxicosis	Swine	Enteritis; emesis
T-2 toxicosis	Swine, cattle, poultry	Dermonecrosis; gastroenteritis
Diacetoxyscirpenoı	Swine	Oral, gastroenteric necrosis, haemorrhage
Leukoencephalo malacia	Horses	Nervous depression, incoordination
F-2 toxicosis (zearalenone)	Swine	Oestrogenism
Ochratoxicosis	Swine, turkeys	Nephropathy
Paspalum staggers	Cattle, sheep, horses	Ataxia
Slaframine toxicosis	Cattle, sheep	Salivation, diarrhoea, polyuria
Stachybotryotoxicosis	Horses	Dermonecrosis; gastroenteritis; haematopoietic depression
Tremorgen intoxication	Cattle, sheep, dog	Fasciculation; ataxia; prostration

Table 4.3. Responses of cattle to various doses of selected mycotoxins (Pier *et al.*, 1980)

Mycotoxin	Dosage	Duration	Effect
Aflatoxin	0.08 mg kg^{-1} [a]	2+ weeks	Reduced weight gain (calves)
	0.2 mg kg^{-1} (B1)	2–4 weeks	Reduced weight gain; coagulopathy (calves)
	0.7 ppm (B1) [b]	19 weeks	Reduced weight gain (steers)
	2 ppm	-	Reduced milk production
	0.5 mg cow^{-1} day^{-1}	-	Detectable milk residue
	14 to 46 ppb (B_1)	-	
	Approx. feed (B_1) milk (M_1) ratio = 200:1	-	
	0.5 mg kg^{-1}	14 days	Death (calf); icterus haemorrhage; hepatic necrosis; coagulopathy
Ochratoxin A	1.0 ppm	59 days	Death (steer)
	1.0 mg kg^{-1}	14 days	Depression; reduced weight gain; nephritis enteritis
T-2 toxin	2.0 mg kg^{-1}	14 days	Coagulopathy (calf)
	0.16 mg kg^{-1}	12 days	Enteritis; abomasal ulcers
	0.64 mg kg^{-1}	20 days	Death (calf); bloody faeces; enteritis; abomasal and ruminal ulcers; coagulopathy

[a] mg kg^{-1} = daily intake of toxin kg^{-1} of body weight.

[b] Feed content expressed as ppm or ppb.

Table 4.4. Effects of mycotoxins on poultry (Pier *et al.*, 1980)

Mycotoxin	Effects	Quantity of toxin (ppm)
Aflatoxin	Acute death, hepatic necrosis and haemorrhage	1–10
	Impaired immunogenesis	0.25
	Reduced resistance	0.6–1.0
	Decreased gain	1.5–2.5
	Decreased egg production	2–8
Ochratoxin	Acute disease, diarrhoea, death	4–16
	Toxic nephropathy	4
	Reduced gain	2–4
	Decreased egg production	2
T-2 toxin	Oral necrosis	4
	Reduced gain	4
	Decreased egg production	20

Table 4.5. Responses of pigs to various amounts of selected mycotoxins (Pier *et al.*, 1980)

Mycotoxins	Effect	Quantity of toxin
Aflatoxin		
20 kg pig	Decreased growth rate	0.26 ppm
20 kg pig	Impaired immunogenesis	0.86 ppm[a]
22 kg pig	Acute fatal toxicosis	2–4 ppm
6.5 kg pig	Single oral dose LD_{50}	0.62 mg kg^{-1} of body weight
Ochratoxin A		
20 to 90 kg pigs	Chronic nephropathy	0.2 ppm
3 kg pig	Nephrosis, slow growth	2.0 ppm[a]
3 kg pig	Acute fatal enteritis	10.0 ppm[a]
Diacetoxyscirpenol		
Growing pigs	Haemorrhagic enteritis	0.38 ppm
Zearalenone		
Growing gilts	Oestrogenism	1–5 ppm

[a] Level estimated from mg kg^{-1} dose and National Research Council Feed Consumption Data.

CHRONIC PRIMARY MYCOTOXICOSES

In these types of mycotoxicoses there is a relative lack of macroscopically visible changes in the infected animals and this prevents an easy diagnosis based on symptoms.

In chronic mycotoxicoses in calves, pigs and poultry the effects appear as reduced productivity in the form of slower growth rates, reduced reproductive efficiency and inferior market quality. Reduced feed conversion efficiency resulting in reduced weight gains and a general lack of thrift in the animals has been associated with aflatoxin, ochratoxin A and T-2 toxins. Reduced milk yields in dairy cows and reduced egg production and increased cracked eggs in poultry has been noted in mature animals exposed to aflatoxin and ochratoxin A (Table 4.6). T-2 toxin can also affect egg production as well as shell thickness and strength. Aflatoxin can cause a generalized loss of tissue strength and integrity which may cause an increased susceptibility to bruising and, with chickens, a decreased market acceptability. Furthermore,

Table 4.6. Responses of animals to chronic aflatoxin B_1 ingestion (Bryden, 1982)

Animal	Dietary concentration ($\mu g\ kg^{-1}$)	Effect
Cattle		
calves	110	Reduced weight gain
steers	700	Reduced weight gain
cows	1500	Reduced milk yield
calves	200	Coagulopathy
Pigs		
20 kg	260	Reduced weight gain
50 kg	385	Reduced weight gain
50 kg	385	Reduced feed intake
50 kg	1480	Reduced feed conversion
20 kg	860	Immunosuppression
sow	500	Reduced feed intake
Poultry		
broiler	500–1000	Reduced weight gain
broiler	500–1000	Reduced feed intake
broiler	500–1000	Reduced feed conversion
poult	250	Reduced weight gain
layer	200–8000	Reduced egg production
broiler	625	Bruising
broiler	2500	Coagulopathy
poult	250	Immunosuppression
broiler	625	Immunosuppression

this problem is only seen after the birds have been killed and prepared for market. Levels of aflatoxin required in feed to cause this effect in broiler chickens can be as low as $0.6\,\mu g\,g^{-1}$ feed which is less than half that required to reduce growth!

Diagnosis will depend mainly upon the absence of other readily diagnosed diseases and the finding of mycotoxins in the suspect feed. Here, again, the problem is compounded by the difficulty in most intensive animal rearing systems to obtain a representative sample of feed, since in most instances animal feed is rarely stored on the farm but is rapidly consumed shortly after delivery. Thus, by the time a mycotoxicosis is suspected, the suspect feed has been consumed.

SECONDARY MYCOTOXIN DISEASES

Consumption of low levels of certain mycotoxins can lead to impairment of the native and acquired resistance to infectious diseases causing health-related economic losses and vaccine failures. To the veterinarian the effects of secondary mycotoxin disease is simply to enhance the infectious processes to which the host animal is naturally predisposed. In particular, low concentration aflatoxin-induced immunologic deficiency is believed to be associated with the specific failure to function of the cell-mediated immune system, while at slightly higher levels, antibody production may also be impaired. Complement and interferon production are decreased as is the activity of lymphocytes and phagocytes. Mycotoxins generally affect the non-specific humoral substances and phagocytotic processes while certain cellular immune mechanisms, involving the function of T lymphocytes and delayed hypersensitivity, have been affected by aflatoxin.

Aflatoxin consumption has been associated with increased susceptibility to candidiasis, Marek's disease, coccidiosis and salmonellosis in chickens, pasteurellosis and salmonellosis in turkeys, erysipelas and salmonellosis in pigs, fascioliasis and clostridial infections in calves and intramammary infections in cows. The levels of toxins causing immunosuppressant effects occurs at much lower levels of intake than those which cause chronic effects such as growth rate reduction (Table 4.6). Administration of trichothecenes to animals induces more susceptibility to microbial infection, while the immunological effects of a simple sublethal dose of trichothecenes seems to be transient *in vivo*.

Ochratoxin A can be an immunosuppressant to mice in sublethal quantities and is considered to act in a non-selective immuno-suppressive manner. In many broiler chickens sublethal dietary levels of ochratoxin A fed over a 20-day period caused a significant depression in immunoglobulin-containing cells in all lymphoid organs studied while total immunoglobulin levels were also reduced in the sera (Fig. 4.2). In field outbreaks of ochratoxicosis in broiler fowls an increased incidence of air sacculitis infections has been observed.

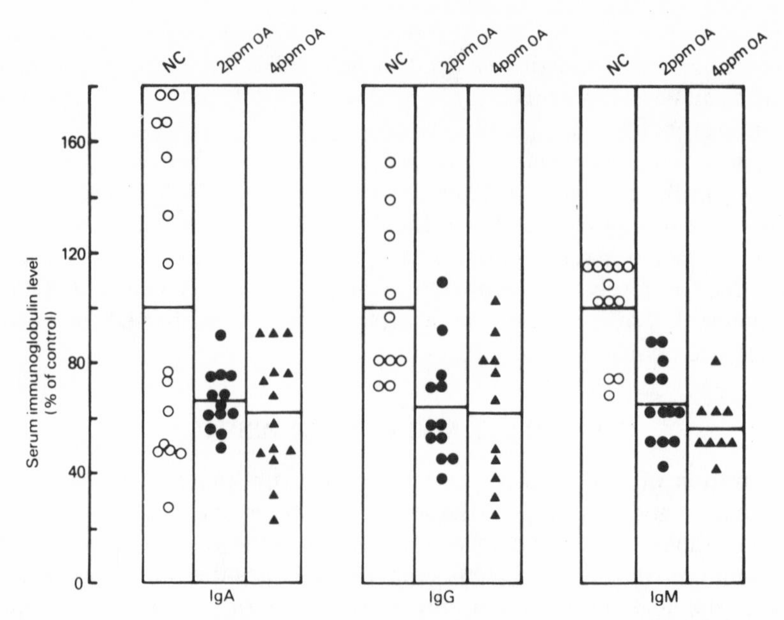

Fig. 4.2. Effect of ochratoxin A on the immunoglobulin levels in the sera of broiler chicks. NC: normal concentration. (Dwivedi and Burns, 1984)

It has now become increasingly apparent that mycotoxin-related animal problems seldom represent a straightforward disease situation. Correct diagnosis will always be difficult. It is quite probable that a mycotoxin, occurring as a single aetiological agent, will rarely be the cause of a mycotoxin-induced disease. More generally, a mycotoxicosis will result not only from the contamination of a feed supply with a mycotoxin but, in addition, there will be other detrimental situations occurring at the same time such as nutritional imbalances or the presence of other toxic molecules including other mycotoxins etc. Consequently, the net result of these multiple diverse situations will give rise to the observed symptoms which will, undoubtedly, be different in varying degrees from the symptoms produced by a single toxic assault. This will create confusion for the veterinarian and quite probably hinder a rapid and accurate diagnosis.

Field-derived mycotoxicoses involving naturally occurring mycotoxins, are now believed to arise from much lower concentrations of mycotoxin than would be required under laboratory administered conditions. This has been determined from the following practical observations.

1. Laboratory studies usually involve small numbers of animals per treatment (10–100) whereas field observations deal with much higher numbers (e.g. with poultry, several thousands). Thus, in laboratory studies, large effects are necessary before accepted statistical analyses will indicate a significant effect. In field episodes slight effects can cause considerable economic loss, e.g. 1 per cent decrease in weight gain is 'highly' significant.

2. Laboratory animals are kept in clearly defined beneficial environmental conditions, whereas field animals may be exposed to nutrient imbalances, subclinical diseases, poor management practices and other stresses such as vaccinations, etc. Since it is well-documented from controlled experiments that mycotoxin response in animals is strongly influenced or accentuated by adverse environmental conditions, it can be expected that field animals will respond more dramatically to mycotoxin exposure than would comparable laboratory animals.
3. In field-derived mycotoxicoses there is seldom an exhaustive analysis to ensure that not more than one mycotoxin is present in a contaminated sample. Furthermore, in determining the concentration of the suspected mycotoxin it can only ultimately be an estimate of the degree of mycotoxin contamination because of the many problems associated with obtaining a truly representative sample (Chapter 7). Thus comparison with laboratory applied single mycotoxin application is not appropriate.

In essence, laboratory-derived mycotoxin data should be used to correlate certain biological activities of mycotoxins to field problems. Thus, the mycotoxicity arising from the application of various levels of mycotoxins to animals should be based on both laboratory-derived data and field observations.

It has become increasingly difficult and almost impossible to establish, with certainty, a 'safe level' of mycotoxins to farm animals since there can be so many interrelating factors controlling the response of animals to mycotoxin presence in their diet.

BIOCHEMICAL EFFECTS OF MYCOTOXINS

The primary effect of mycotoxins on animals will occur at the cell or molecular level. Most mycotoxins have been extensively examined for their influence on biochemical reactions within the cell, in particular with respect to energy metabolism, carbohydrate and lipid metabolism, protein synthesis and the synthesis and expression of DNA and RNA (Hayes, 1980; Ueno, 1983).

Energy metabolism

Several mycotoxins including aflatoxin B_1, G_1 and M_1, rubratoxin B, patulin and ochratoxin A inhibit oxygen uptake in whole tissue homogenates from several animal species. Ochratoxin A appears to act as a competitive inhibitor of mitochondrial transport carrier proteins while patulin inhibits the enzyme succinic dehydrogenase. Rubratoxin B and aflatoxin B_1 act on the electron transport system. Adenosine triphosphatase activity is inhibited to varying degrees by aflatoxin B_1, rubratoxin B, penicillic acid and patulin.

Carbohydrate and lipid metabolism

With several animal species, reduction of hepatic glycogen levels has been recorded after exposure to aflatoxin B_1, rubratoxin B, penitrem A, ochratoxin A and citrinin. These changes may arise from the effect of the mycotoxins on the synthetic enzymes or by the inhibition of glycogenesis, depression of glucose transport into hepatocytes and acceleration of glycogenesis. Several mycotoxins cause accumulation of hepatic lipids. In chickens, aflatoxin B_1, while affecting lipid synthesis and transport, appears also to influence lipid absorption and degradation.

Nucleic acid and protein synthesis

Although many mycotoxins, including patulin, aflatoxin B_1, T-2 toxin, rubratoxin B and several others, can have pronounced effects on nucleic acid and protein synthesis in several animal species, exact mechanisms of action have not yet been clearly worked out. Aflatoxin B_1 has been variably reported to affect protein synthesis at the level of DNA-dependent RNA polymerase, to affect the membrane and not the polysomes, to inhibit nucleolar RNA synthesis and DNA template function.

Aflatoxin B_1 (most probably as the highly active intermediate 2,3-epoxy-aflatoxin B_1) (Fig. 5.2) can bind to DNA causing disruption of transcription while binding to RNA causes inhibition to protein synthesis. Another active intermediate can form Schiff bases with the free amino groups of functional proteins resulting in reduced enzyme activity.

The trichothecene mycotoxins exhibit potent cytotoxicity to eukaryotic cells—a trait which has been well used in bioassay detection methods. Trichothecene mycotoxins display a high affinity to the ribosomes of eukaryotic cells, especially to 60S subunits of the ribosomes. Although the trichothecenes are also potent inhibitors of DNA and RNA synthesis in whole cells, inhibitor concentration studies indicate that inhibition of protein synthesis is the primary target of these mycotoxins.

BIOLOGICAL EFFECTS OF MYCOTOXINS

Because of their diversity of chemical structures, mycotoxins exhibit a wide array of biological effects and individual mycotoxins can be mutagenic, carcinogenic, embryotoxic, teratogenic or oestrogenic. These effects vary from species to species and cannot always be predicted on the basis of present knowledge derived from animal models or from structure–activity relationships (Hayes, 1980). Biological actions of mycotoxins will be further influenced by sex of animal, environmental factors, nutritional status and interactions with other chemicals. Most mycotoxins primarily damage specific organs and tissues and have been classified as hepatotoxins, nephrotoxins, neurotoxins etc. However, although they mostly show some level of specificity, individual mycotoxins can influence many tissues.

Carcinogenesis

Several mycotoxins have been shown to cause cancers in a variety of animal species. The aflatoxins were the first mycotoxins to be extensively studied for their carcinogenic effect and are now recognized as potent hepatocarcinogens.

Aflatoxins are carcinogenic to mice, fish, rats, marmosets, ducks, tree shrews and monkeys. Oral application is particularly successful but other routes are also effective and the cancers produced are mainly in the liver, colon and kidney. Aflatoxin B_1 primarily causes hepatocellular carcinoma and cholangio carcinoma in the liver. The variation in susceptibility to aflatoxins is well-documented (Table 4.7). The rat has been the most widely used experimental animal and the dose response of a susceptible strain is shown in Table 4.8. The response of the rats to aflatoxin was dose-related up to 100 $\mu g\ kg^{-1}$ with cancer occurring in all animals surviving 18 months. In some instances single dose application followed by normal feeding has resulted in tumour formation in rats. The response to aflatoxin exposure

Table 4.7. Carcinogenicity of aflatoxin (Linsell, 1982)

Species	Dose	Duration of observation	Tumour frequency
Duck	30 $\mu g\ kg^{-1}$ in diet	14 months	8/11 (72%)
Trout	8 $\mu g\ kg^{-1}$ in diet	1 year	27/65 (40%)
Tree shrew	24–66 mg total	3 years	9/12 (75%)
Marmoset	5.0 mg total	2 years	2/3 (65%)
Monkey	100–800 mg total	Over 2 years	3/42 (7%)
Rat	100 $\mu g\ kg^{-1}$ in diet	54–88 weeks	28/28 (100%)
Mice	150 $\mu g\ kg^{-1}$ in diet	80 weeks	0/60 (0%)

Table 4.8. Dose-response in aflatoxin B_1 in male Fischer rats (Linsell, 1982)

Dietary levels ($\mu g\ kg^{-1}$)	Duration (weeks)	Liver cancer frequency
0	74–109	0/18
1	78–105	2/22
5	65–93	1/22
15	69–96	4/22
50	71–97	20/25
100	54–88	28/28

can be influenced by age of animal at exposure while there can be considerable variation between species. It is interesting to note that with the rat females are more resistant than males to both toxic and carcinogenic effects. Nutrient imbalances can predispose animals to cancer development.

Aflatoxin B_1 is the main hepatocarcinogen but G_1 and B_2 have been shown to cause cancers but with reduced potency. The potency of aflatoxin M_1, the milk aflatoxin, to induce liver tumours in trout is about 40 per cent of that of aflatoxin B_1.

The toxicological effects of the aflatoxins only occur after the metabolic activation of the molecules by a microsomal, mixed-function oxidase system. During the complex enzymatic reactions accompanying metabolism and detoxification of aflatoxin B_1, a highly reactive intermediate, 2,3-epoxy-aflatoxin B_1 is generated and this reactive molecule binds covalently with various nucleophilic centres in cellular macromolecules such as DNA, RNA and protein, disrupting their normal biological function. It is believed that

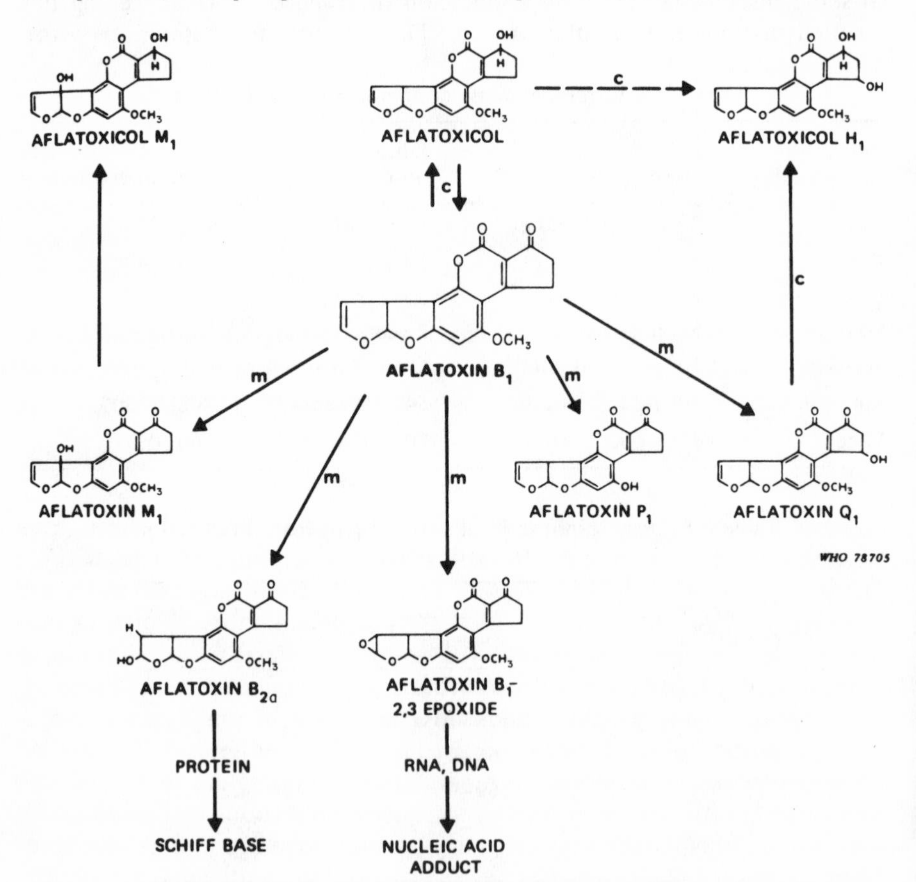

Fig. 4.3. Aflatoxin B_1 metabolism in the liver

the reactive intermediate binds to DNA, disrupting transcription and leading to mutagenesis or carcinogenesis. Binding also occurs to proteins causing reduced enzyme function and toxicity. The toxicity of aflatoxin B_1 can now be considered to result from accumulated effects of the rate of formation of the reactive intermediate, the rate of detoxification of the reactive intermediate and the rate of reaction between the reactive intermediate and target molecules, viz. DNA, RNA and proteins (Fig. 4.3) (see also Chapter 5).

Several other mycotoxins have been shown to induce tumour formation in experimental animals, e.g. sterigmatocystin, penicillic acid, patulin, luteoskyrin, rugulosin, citrinin and ochratoxin A. Most other mycotoxins including zearalenone and the trichothecenes do not appear to be carcinogenic.

The carcinogenic effect of mycotoxins on farm animals is not of much significance since in most cases the useful life span of the animal is normally shorter than that required to produce cancers, i.e. it is marketed before carcinogenic effects are manifest.

Mutagenicity

By using sensitive microbial mutagenic assay systems (Ames test) several mycotoxins have been shown to be mutagenic. Aflatoxin B_1 is the most potent mutagen of the aflatoxins and a strong parallel exists between the ability of the aflatoxins to be mutagenic and carcinogenic. Microsomal activation is an absolute necessity for mutagenicity with aflatoxin. Aflatoxin B_1 can also cause chromosomal aberrations and DNA breakage in plant and animal cells.

Variable levels of mutagenicity have been shown for sterigmatocystin, citrinin, zearalenone, penicillic acid and patulin, while major mycotoxins such as ochratoxin A and T-2 toxin do not seem to be mutagenic.

Teratogenicity

Many mycotoxins are embryotoxic and teratogenic. Prenatal effects have been documented with experimental animals which suggest that aflatoxin B_1, ochratoxin A, rubratoxin B, T-2 toxin and certain ergot alkaloids are teratogenic. Thus, mycotoxins which are potent inhibitors of protein synthesis in eukaryotes must be expected to impair differentiation in sensitive primordia. With hamsters a single intraperitoneal injection of aflatoxin B_1 at 4 mg kg^{-1} body weight, administered on day 8 of pregnancy caused a high proportion of malformed and dead or reabsorbed fetuses. T-2 applied to pregnant female mice caused gross malformation of the limbs and tail, and other malformations to the fetus. Earlier suggestions that zearalenone was responsible for stillbirths and splayleg-malformations in pigs have since been discounted. Recent studies have shown that unidentified *Fusarium* toxins can cause a reduction in hatchability of poultry eggs, i.e. they do not have any influence on egg production but can kill the embryo. Further evidence suggests that these same toxins may also be responsible for tibial dyschondroplasia or bone disorders found in many animals. Most mycotoxins have not been thoroughly screened for teratogenicity.

Oestrogenism

Several animal species will exhibit enlargement of the vulva and uterus when exposed to zearalenone. The pig is the most sensitive domestic species. In the pig the effects of zearalenone are pronounced in the female with swollen vulva, prolapsed rectum, enlarged nipples, and atrophied ovaries while in the male there can be enlarged glands and reduced testes. The effects of zearalenone on the uterus weight in young gilts is shown in Table 4.9. The physiologically active levels in feeds can be as low as 1–5 ppm. The oestrogenic effect of zearalenone can also result in infertility in pigs, rats, mice and guinea pigs and possible susceptibility in cattle, turkeys and chickens.

Table 4.9. Effect of F-2 and *Fusarium*-invaded corn on the uterus of gilts[a] (Mirocha, 1984)

Treatment	Weight of gilt[b] (kg)	Weight of uterine horn (g)	
		Total	Per kg body weight
Control, sacrificed after 5 days	34.1	27.8	0.81
Control, sacrificed after 14 days	75.5	28.0	0.37
Oestradiol, 0.5 mg IM daily for 4 days	60.5	173.1	2.86
Oestradiol, O.5 mg IM daily for 14 days	81.4	241.6	2.97
F-2, 1 mg daily for 8 days	49.5	47.0	0.95
F-2, 5 mg daily for 8 days	57.7	85.4	1.48
F-2, 10 mg daily for 5 days	53.6	174.6	3.26
F-2, 25 mg daily for 5 days	76.3	271.6	3.56
F-2, 50 mg daily for 4 days	78.6	251.9	3.20
Fusarium-invaded corn, 80 g daily for 6 days	59.5	136.9	2.30
Fusarium-invaded corn, 80 g daily for 14 days	75.5	206.0	2.73

[a] After administration *per os*.

[b] The proprietary material oestradiol (ECP, Upjohn Co.) was administered by intramuscular injection (IM); F-2 by gelatin capsule; *Fusarium*-invaded feed was mixed with normal ration. Prepubertal pigs about 6 weeks of age.

The biological effects of most mycotoxins mainly occur through specific organs and/or tissues and consequently mycotoxins can be classified as hepatotoxins, nephrotoxins, neurotoxins or dermal toxins. While a mycotoxin may cause primary lesions in a target organ damage to other tissues and organs may also occur.

Hepatotoxins

The hepatic tissues of the liver can absorb toxic substances from the bloodstream and thus remove them from circulation. It must be anticipated that toxic molecules such as the mycotoxins once inside the animal system may well reach the liver where they can be detoxified and excreted. Several mycotoxins, including the aflatoxins, rubratoxin B, the sporidesmins and citrinin induce both non-specific liver injury such as fatty and pale livers, moderate to extensive necrosis and haemorrhage, together with specific lesions.

Aflatoxin B_1 is certainly the most documented and studied hepatotoxic mycotoxin. The susceptibility of animals to aflatoxin B_1 varies but in all cases the primary cause of illness and death results from liver damage. Liver injury has been demonstrated both in field outbreaks of aflatoxicoses and in laboratory-administered conditions with poultry, pigs, cattle and dogs. In feeding experiments, farm animals were fed on diets containing aflatoxins (0.3 to several mg kg^{-1}) for periods of time ranging from a few weeks to a few months and then sacrificed and subjected to post-mortem pathology. Liver lesions were consistently observed with most toxin levels, particularly at LD_{50} toxin levels. Similarly, liver lesions have been observed when aflatoxins have been administered to laboratory animals such as rat, cat, guinea pig, monkey and rabbit. The hepatic lesions induced by aflatoxin B_1 in the duckling have become the basis of a bioassay and at sublethal doses of the toxins, toxicity values are expressed on the degree of biliary proliferation.

In cases of acute aflatoxin toxicity the liver of the animal appears pale, decolourized and generally increases in size. In ducklings, turkeys and young pigs, necrosis of the hepatic parenchyma of the liver and haemorrhages are regularly observed. If death has not occurred rapidly, there is a characteristic proliferation of undifferentiated cells in the portal spaces of the liver. The principal symptoms of aflatoxin liver damage in a variety of animals is shown in Table 4.10. Liver damage is normally accompanied by derangement of normal blood clotting mechanisms, reduction in total serum proteins and increases in certain serum enzymes.

In chronic aflatoxin toxicity the liver is congested and shows haemorrhagic and necrotic zones, with proliferation of epithelial cells of the bile ducts in cattle, turkeys and ducklings; with chickens there is destruction of the cells of the hepatic parenchyma and proliferation of the epithelial cells during the first few weeks of exposure followed by an infiltration of lymph cells. With prolonged exposure of low concentrations of aflatoxin, cancer of the liver may occur.

Table 4.10. The principal symptoms of aflatoxicosis among a variety of animals (+ presence of a symptom, − absence) (Moreau, 1979)

	Cow	Pig	Sheep	Duckling	Turkey	Chicken
Lesions of liver						
acute necrosis and haemorrhage	−	−	−	+	+	−
chronic fibrosis	+	+	−	−	−	−
regenerative nodules	+	+	−	−	+	−
bile duct proliferation	+	+	−	+	+	+
vascular occlusions	+	−	−	−	−	−
Hepatic cells						
megalocytosis	+	+	−	+	+	−
giant nuclei	+	+	−	+	+	+
infiltration of mast cells	−	−	−	−	+	+

Ochratoxin A intoxications, although primarily nephrotoxic in nature, can also cause damage to the liver of the animal. In recent studies with young broiler chicks on sublethal levels of intake of ochratoxin A, liver damage also accompanied nephrotoxicity. In the liver ochratoxin A caused lymphoid cell infiltration and glycogen accumulation in a dose-related response. There was regression of the lymphoid organs and a major reduction of lymphoid cell population.

Apart from kidney damage, the aflatoxins in particular can also cause bilateral adrenal haemorrhages and petechial haemorrhages in many organs such as in congested lungs, myocardium and spleen. Extensive haemorrhage symptoms have been demonstrated in guinea pigs, cattle and chickens. Many mycotoxins can cause lesions in animal tissues which are often accompanied by haemorrhages. Such haemorrhages are associated with varying degrees of fragility of blood capillaries either on the surface of skin or on organs such as liver, kidneys, lung and brain. Haemorrhages of the intestinal tract result in bloody diarrhoea.

Nephrotoxicity

The kidney is the major organ involved in the excretion or elimination of waste products and foreign substances that are not utilized by the body. What evidence exists to relate mycotoxin intake with renal malfunction in animals? Several mycotoxins including ochratoxin A, citrinin, aflatoxin B_1, rubratoxin B and sterigmatocystin can induce variable levels of renal damage both in field intoxications and in experimentally dosed animals. Ochratoxin A and citrinin have been extensively studied.

Porcine mycotoxic nephropathy has been well documented in Denmark for many years and the disease may well be present in other European countries and North America. The main disease determinant of porcine nephropathy is a consistent presence of ochratoxin A in the feed and the regular occurrence of the mycotoxin in the kidneys of slaughtered animals (2–70 $\mu g\ kg^{-1}$). Ochratoxin A has never been detected in healthy animal kidneys. Examination of the diseased kidneys shows degeneration of the proximal tubules, followed by atrophy of the tubular epithelium, interstitial fibrosis in the renal cortex and hyalinization of some glomeruli.

The typical disease symptoms found in field cases of mycotoxic porcine nephropathy have been reproduced in experimental animals fed on diets containing levels of ochratoxin A normally associated with natural outbreaks of this disease. Pigs fed rations containing 200–4000 $\mu g\ kg^{-1}$ showed nephropathy symptoms over four months and chickens and hens fed 300–1000 $\mu g\ kg^{-1}$ for one year developed avian nephropathy. Ochratoxin A has also been shown to produce nephropathy in a controlled, laboratory experiment in rats. Although rats will die rapidly with high doses of ochratoxin A obvious effects on kidney function were not noted until small, repeated doses were administered. In these cases there was a significant depression of urine osmolarity and increased excretion of protein and glucose indicating disruption of renal function. This and other evidence suggested that ochratoxin A acted on the proximal tubules of the kidney.

Recent studies with broiler chickens fed sublethal doses of ochratoxin A (2 and 4 ppm) for 20 days from hatch caused significant enlargement of the kidney, liver and proventriculus whereas the thymus and Bursa of Fabricius were reduced in size. The main effect on the kidney was on the proximal convoluted tubules causing severe distension, enlargement and hypertrophy together with thickening of the glomerular basement membrane. Considerable ultrastructural changes have been observed in the kidney of broiler chickens, dogs, pigs, mice, rats and ducklings when fed on diets containing nephrotoxic levels of ochratoxin A.

Although ochratoxin A can cause liver damage, the main effect of toxin consumption is nephrotoxic; similarly, the effect of aflatoxin in most animals is more hepatotoxic than nephrotoxic.

Citrinin causes distinct renal damage when administered to experimental animals such as rabbits, guinea pigs, rats and pigs causing swelling of the kidneys and acute tubular necrosis.

Neurotoxicity

Several mycotoxins characterized as tremorgens have been isolated from various fungi. At least 20 tremorgenic mycotoxins are now recognised including the ergot tremorgens isolated from *Claviceps paspali* and others from several *Penicillium* and *Aspergillus* spp. commonly found in soil. Since tremorgens are more consistently present on natural forage, ruminants are generally the animals most likely to ingest sufficient quantities to cause neurological disorders. However, other animals can show similar symptoms following ingestion of toxins. Most tremorgens are indole derivatives.

Tremorgenic toxins are now generally believed to be the causal agents in certain idiopathic disorders of sheep and cattle, such as ryegrass staggers, Bermuda grass staggers and marsh staggers in New Zealand, USA and Britain. Stationary animals show tremors in skeletal muscles and head tremor, and when walking, the animals become uncoordinated, moving with a rocking horse action with extended front legs. Animals may collapse but mortality is rare.

Experimental production of a staggers syndrome with repeated dosing with small quantities of mycelium containing penitrems (tremorgenic compounds), has been demonstrated in several research studies, and has further confirmed the similarity between the natural syndrome and experimentally induced syndrome. Animals may receive tremorgenic mycotoxins by consuming infected grasses while recent studies also suggest that tremorgens may be produced in the soil environment and translocated into the leaves of grasses and legumes by way of the roots. The exact mechanism of action of the tremorgenic mycotoxins implicated in the 'staggers' syndrome causing tremor and incoordination is not yet known but certainly the normal pattern of amino acid neurotransmitter release is disturbed.

Citreoviridin and patulin have also been considered to influence the nervous systems of affected animals. Citreoviridin causes convulsions and paralysis, respiratory and cardiac failure and with high dosages, eventual death. Patulin causes vomiting in experimental animals and roquefortine can produce convulsive seizures in mice.

Vomiting and refusal of feeds are important symptoms of trichothecene toxicoses in ducks, cats, dogs and pigs. It has been considered that these toxins may stimulate the chemoreceptor trigger zone in the medulla oblongata.

Dermal toxicosis

The trichothecenes, when applied to the skin of animals, can cause local irritation, inflammation, desquammation, subepidermal haemorrhaging and general necrosis. Topical application of suspected trichothecene-containing samples to shaved animal skin is a well-used bioassay method. Oral administration may also cause listlessness and inactivity, diarrhoea and rectal haemorrhage. With toxic levels, the mucosal epithelium of the stomach and small intestine erode with accompanying haemorrhaging and eventual death of the animal.

In broiler chickens necrotic oral lesions have been produced when fed a diet containing purified T-2 toxin. The degree of oral necrosis was dose-related and it has been concluded that this is the primary effect of experimental T-2 toxicosis in this animal. Diacetoxyscirpenol causes more severe oral necrosis than T-2 toxin. In field outbreaks of oral necrosis, it is still not clear which trichothecene is the primary cause. T-2 toxicosis in laying hens is also characterized by an oral response similar to that observed with chickens. Parallel symptoms perhaps resulting from feeding problems arising from the oral necrosis include marked decline in feed intake and body weight as well as egg production.

A haemorrhagic disease syndrome is one of the most frequently described symptoms in animal mycotoxicoses. Such haemorrhages are associated with fragility of the blood capillaries and may occur on the surface of the skin or in various organs such as the gastrointestinal tract, liver, kidneys, adrenal glands, lungs, brain, etc. However, although experimentally applied trichothecenes can cause haemorrhages there is little conclusive evidence to support a suggested association between these trichothecene mycotoxins and the occurrence of haemorrhagic disease in farm animals. Chronic oral doses of diacetoxyscirpenol, T-2 toxin, crude extracts or whole cultures of *Fusarium tricinctum* to calves, cattle and pigs have given negative results, further emphasizing the minor aetiological roles of these toxins in feed-associated haemorrhagic disease syndromes.

Several mycotoxins including the furocoumarins, butenolide and aflatoxin B_1 can cause dermatoxic effects when skin is exposed to the toxin and sunlight simultaneously. The sporidesmins, the causal agents of facial eczema and oedema of grazing animals, are believed to exert their effect by causing the release of photosensitising agents into the bloodstream, which then give rise to eczema and dermatoses.

INTERACTION OF NUTRITION AND MYCOTOXINS

The achievement of optimum health and disease resistance in an animal is a function of nutrient intake and nutrient status. Nutritional imbalances have long been known to occur in the feed industry and in many cases these imbalances result in disease symptoms not unlike those resulting from mycotoxicoses. For the veterinarian they pose serious problems in diagnosis and remedy.

However, in recent years, there have been many experimental attempts to classify the role of nutritional factors and presence of mycotoxins. The relationship of nutrition imbalances to mycotoxicoses has been extensively examined with the aflatoxins. Aflatoxicosis has been shown to interreact with protein, lipid and vitamin metabolism.

Thus it is well recognized that protein deficiencies exacerbate the effects of aflatoxins in chickens, rats and pigs while diets supplemented with protein show a protective effect with chickens. Broiler chickens fed on a sublethal level of aflatoxins showed lower levels of amino acids in the plasma compared with control chickens. The degree of amino acid reduction varied from one amino acid to another, but particularly the sulphur-containing amino acids, e.g. methionine, showed the greatest reduction. Methionine is of particular significance in poultry nutrition and studies with a normal diet (A), a diet slightly deficient in methionine (B) and finally a diet fortified with methionine (C) showed that normal growth was achieved with diet A and C but not with B. However, in the presence of aflatoxin, birds on diet B showed further reduced growth, birds on diet A showed some growth reduction whereas birds on diet C showed no growth reduction. Thus, in these experiments, increasing the methionine level in the diet had a sparing effect upon the growth depressing effects of aflatoxin. It has been suggested that perhaps intestinal absorption plays a major role during aflatoxicosis and that this may be particularly overcome by dietary fortification.

It has also been suggested that the sulphur-containing amino acids such as cysteine and methionine alter the biochemical detoxification of aflatoxin. Glutathione, a sulphur-containing tripeptide, is now believed to bind the active epoxide aflatoxin molecule in the liver thus rendering it non-toxic and able to be excreted in the bile. Extra dietary sulphur-containing amino acids alleviate the growth inhibiting effects of dietary aflatoxin, possibly by increasing glutathione concentration. Thus, extra dietary protein may offset the effect of aflatoxin by increasing the concentrations of the precursors of glutathione.

Lipid metabolism, as judged by increased content of lipid in the liver, is altered during aflatoxicosis in pigs, cattle, chickens, turkeys, ducks, dogs, cats, monkeys and man. Other mycotoxins including ochratoxin A and citrinin also affect lipid metabolism. What is so important about a fatty liver—a symptom common to several types of mycotoxicoses? This arises since, depending on the species, from 40 to 95 per cent of the total lipid biosynthesis in animals occurs in the liver and is then transported to the sites where it is utilized. In this way a fatty liver can be a sensitive indicator of a disturbance in lipid metabolism. Thus, an effect on lipid metabolism is an early if not primary effect of aflatoxicosis. Studies with chickens have even shown that aflatoxin concentrations too small to affect growth rate alters lipid metabolism. Aflatoxin may affect the formation of enzymes involved in lipid metabolism. Aflatoxin B_1 not only affects lipid synthesis and transport but also interferes with lipid absorption and degradation in chickens.

A topic which bridges lipid and vitamin metabolism during aflatoxicosis is the effect of lipotrope-deficient diets on the expression of aflatoxicosis. Lipotropes are compounds such as methionine, vitamin B_{12}, folic acid and choline which acquired their generic name because their presence in a diet prevents a fatty liver. A diet marginally deficient in vitamin B_{12} and choline enhanced the development of tumours in rats whereas a severely deficient diet did not affect aflatoxin carcinogenicity.

Dietary deficiencies of riboflavin and vitamin D_3 increased the sensitivity of chickens to aflatoxin whereas a thiamine deficiency has a protective effect. Dietary supplementation with choline, inositol and vitamins B_{12}, E and K had no beneficial effect in chickens subjected to low aflatoxin levels in their diet. Aflatoxin interferes with metabolism of B vitamins in chickens. Liver fatty acid composition is also severely influenced by aflatoxin B_1 but the extent of these changes could be reduced by the inclusion of additional dietary biotin.

Vitamin A content of liver is reduced by aflatoxin in pigs, cattle and chickens. This implies a deficiency of vitamin A which might be corrected by dietary amendment to the benefit of animals with aflatoxicosis. Although several feeding trials with vitamin additions have been inconclusive, several studies have shown that vitamin A deficient diets enhance the cancer-inducing effect of aflatoxin in rats. The case for the involvement of vitamins D, E and K in aflatoxicosis therapy is uncertain at best.

It is undoubtedly true that the nutrient status of the diets being ingested by an animal can markedly affect the expression of mycotoxin toxicity. Furthermore, mycotoxins also alter nutrient metabolism and impose a further nutritional stress in that reduced feed intake is an early symptom of many mycotoxicoses.

FURTHER READING

Anon (1979). Mycotoxins, *Environmental Health Criteria, 11.* World Health Organisation, Geneva.

Bababunmi, E. A. and Bassir, O. (1982). A delay in blood clotting of chickens and ducks induced by aflatoxin treatment. *Poultry Science*, **61**, 166–168.

Bryden, W. L. (1982). Aflatoxins and animal production: an Australian perspective. *Food Technology in Australia*, **34**, 216–223.

Bryden, W. L., Cumming, R. B. and Balnave, D. (1979). The influence of vitamin A status on the response of chickens to aflatoxin B, and changes in liver lipid metabolism associated with aflatoxicosis. *British Journal of Nutrition*, **41**, 529–541.

Burditt, S. J., Hagler (Jr), W. M. and Hamilton, P. B. (1983). Survey of molds and mycotoxins for their ability to cause feed refusal in chickens. *Poultry Science*, **62**, 2182–2191.

Ciegler, A., Kadis, S. and Ajl, S. (Eds.) (1971). *Microbial Toxins*, vols. 6, 7, 8, Academic Press, New York.

Chang, K., Kurtz, H. and Mirocha, C. J. (1979). Effects of the mycotoxin zearalenone on swine. *American Journal of Veterinary Research*, **40**, 1260–1267.

Davis, N. D. and Diener, U. L. (1978). Mycotoxins. In: *Food and Beverage Mycology*, L. R. Beuchat (Ed.), Avi Publishing, Connecticut, pp. 397–444.

Doerr, J. A., Huff, W. E., Wabeck, C. J., Chaloupka, G. W., May, J. O. and Merkley, J. W. (1983). Effects of low level chronic aflatoxicosis in broiler chickens. *Poultry Science*, **62**, 1971–1977.

Divivedi, P. and Burns, R. B. (1984). Pathology of ochratoxicosis A in young broiler chicks. *Research in Veterinary Science*, **36**, 92–103.

Divivedi, P. and Burns, R. B. (1984). Effect of ochratoxin A on immunoglobulins in broiler chicks. *Research in Veterinary, Science*, **36**, 117–121.

Hamilton, P. B. (1977). Lipid and vitamin metabolism during mycotoxicoses. *Federation Proceedings*, **36**, 1899–1902.

Hamilton, P. B. (1977). Interrelationships of mycotoxins with nutrition. *Federation Proceedings*, **36**, 1899–1902.

Hayes, A. W. (1980). Mycotoxins: Biological effects and their role in human diseases. *Clinical Toxicology*, **17**, 45–83.

Krogh, P. (1976). Mycotic nephropathy. In: *Advances in Veterinary Science and Comparative Medicine*, vol. 20, C.A. Brandly and C.E. Cornelius (Eds), Academic

Press, New York, pp. 147–170.
Krogh, P. (1977). Ochratoxins. In: *Mycotoxins in Human and Animal Health*, J. V. Rodricks, C. W. Hesseltine and M. A. Mehlman (Eds), Pathotox Illinois, pp. 489–498.
Lanza, G. M., Washburn, K. W. and Wyatt, R. D. (1982). Strain variation in hematological response of broilers to dietary aflatoxin. *Poultry Science*, **60**, 500–504.
Linsell, A. (1982). Carcinogenicity of mycotoxins. In: *Environmental Carcinogens Selected Methods of Analysis*, vol. 5, *Some Mycotoxins*, H. Egan, L. Stoloff, M. Castegnaro, P. Scott, I. K. O'Neill and H. Bartsch (Eds), International Agency for Research on Cancer, IIARC Scientific Publ;ications, Lyon, pp. 3–14.
Mirocha, C. J. (1984). Mycotoxicoses associated with Fusarium. In: *The Applied Mycology of Fusarium*, M. O. Moss and J. E. Smith (Eds), British Mycological Symposium. vol. 7, pp. 141–155.
Moreau, C. (1979). *Moulds, Toxins and Food*, Wiley, New York.
Morehouse, L. G. (1979). Mycotoxicoses of the bovine with reference to fungi and toxins associated with disease. *The Bovine Practitioner*, **14**, 175–180.
Neal, G. (1984). Involvement of mycotoxins in acute and chronic animal disease—metabolic aspects. *Chemistry and Industry*, 542–546.
Newborne, P. M. and Butler, W. H. (1969). Acute and chronic effects of aflatoxin on the liver of domestic laboratory animals: a review. *Cancer Research*, **29**, 236–250.
Ong, T. (1975). Aflatoxin mutagenesis. *Mutation Research*, **32**, 236–250.
Pathre, S. V. and Mirocha, C. J. (1979). Trichothecenes: natural occurrence and potential hazard. *Journal American Oil Chemists Society*, **86**, 820–823.
Pier, A. C., Richard, J. L. and Thurston, J. R. (1978). The influence of mycotoxins on resistance and immunity. In: *Interaction of Mycotoxins in Animal Production*, National Academy of Science, Washington DC, pp. 56-66.
Pier, A. C., Richard, J. L. and Cyzewski, S. J. (1980). Implications of mycotoxins in animal disease. *Journal American Veterinary Medicine Association*, **176**, 719–724.
Purchase, I. F. H. (Ed.) (1974). *Mycotoxins*, Elsevier, New York.
Spensley, P. C. (1963). Aflatoxin, the active principle in turkey 'X' disease. *Endeavour*, **22**, 75–79.
Ueno, Y. (Ed.) (1983). *Trichothecenes - Chemical, Biological and Toxicological Aspects*, Elsevier, Amsterdam, New York.
Vesonder, R. F. and Hesseltine, C. W. (1981). Vomitoxin: natural occurrence on cereal grains and significance as a refusal and emetic factor to swine. *Process Biochemistry*, **16**, 12–14.
Voight, M. N., Wyatt, R. D., Ayers, J. C. and Koehler, P. (1980). Abnormal concentrations of B vitamins and amino acids in plasma, bile and liver of chicks with aflatoxicosis. *Applied and Environmental Microbiology*, **40**, 870–875.
Wylie, T. D. and Morehouse, L. G. (1977) (Eds) Mycotoxic Fungi, Mycotoxins and Mycotoxicoses. *Encyclopaedic Handbook*, vols. 1–3, Marcel Dekker, New York.
Wogan, G. N. and Newborne, P. M. (1967). Dose response characteristics of aflatoxin B1 carcinogenesis in the rat. *Cancer Research*, **27**, 2370–2376.

CHAPTER 5

Human Mycotoxicoses

Although the poisonous nature of some toadstools had been appreciated for many centuries, it was not until the mid-nineteenth century that a microfungus, *Claviceps purpurea*, could be directly associated with disease in man. the disease itself, ergotism, had been well-documented since the middle ages but its aetiology remained a mystery. Today, the activity of the ergot alkaloids is well understood. They interfere with certain forms of nerve activity; they can cause the constriction of peripheral blood capillaries which can lead to oxygen starvation and gangrene of the limbs; they stimulate smooth muscle activity and may also influence central nervous activity.

The long time lag between the description of ergotism and an understanding of its aetiology could be considered as a reflection of the gradual development of the essential scientific disciplines of mycology and chemistry. However, even since these disciplines, and others such as biochemistry and toxicology, had fully emerged, there have still been delays in associating moulds and their toxins with diseases in man. The detailed study of mycotoxins and mycotoxicoses is really a phenomenon of the second half of this century.

There is no doubt now about the role of mycotoxins in ergotism, alimentary toxic aleukia, yellow rice disease and acute aflatoxicosis of man. There are other diseases with a puzzling aetiology for which it is possible that mycotoxins play a significant role but which remain a mystery. Thus, despite the suggestion by Dr Barnes as long ago as 1967 that Balkan nephropathy in man may be associated with mycotoxins, such a link has still not been proven (Anon, 1979). Nevertheless, 20 years of research have not implicated any other aetiological agents such as bacteria, viruses, toxic metals or genetic factors, and two observations do seem to implicate mycotoxins. A strong correlation has been shown between the number of people dying of nephropathy and excess rainfall during the two previous years (Austwick, 1975), and such patterns of rainfall may be associated with seasonal moulding of foods, and the symptoms do resemble those of porcine nephropathy caused by the mycotoxin ochratoxin A (Krogh *et al.*, 1977).

AFLATOXIN AND LIVER CANCER

Since the recognition in the 1960s that aflatoxin can induce liver carcinomas in animals as diverse as the rainbow trout and the rat, studies of the epidemiology of liver cancer in man have included consideration of the possibility that aflatoxin may be an aetiological agent in this disease. The hypothesis that mycotoxins have a role in the aetiology of liver cancer in man was clearly suggested by Oettle in 1965 and was hinted at even earlier by Le Breton *et al.* in 1962. In at least three parts of the world, Thailand, The Philippines and East Africa, good epidemiological evidence has been collected showing a correlation between the incidence of liver cancer and exposure to aflatoxins. Table 5.1 shows a comparison of the incidence and concentration of aflatoxin in groundnut products consumed in these countries with that in the USA and Canada where the incidence of liver cancer is low.

Table 5.1. Incidence and concentration of aflatoxins in peanut products for human consumption in various countries. (Taken from an excellent review by Stoloff, 1977)

Country	Period of analysis	No. of samples	Aflatoxin B_1 % incidence	Aflatoxin B_1 Mean conc. $\mu g\ kg^{-1}$
USA and Canada	1972–5	1416	19	1
Thailand	1967–9	216	54	470
Uganda	1966–7	150	19	70
Philippines	1967–9	309	88	130

The studies in East Africa conducted by Peers and Linsell (1977) were so sufficiently detailed that it was possible to demonstrate that men are apparently more sensitive than women to increasing exposure to aflatoxin (Fig. 5.1). It is known from other studies that in mammals such as the rat, the male is more sensitive to the carcinogenic activity of aflatoxin than the female (Table 5.2).

Table 5.2. Influence of sex on the induction of liver tumours in the rat by aflatoxin B_1. (Data from Newberne and Wogan, 1968)

Conc. aflatoxin B_1 in diet (mg kg^{-1})	Average time of tumour formation days: male	Average time of tumour formation days: female	Total intake of aflatoxin (mg per rat): male	Total intake of aflatoxin (mg per rat): female
1.0	245	448	2.9	5.9
0.015	476	560	0.095	0.115

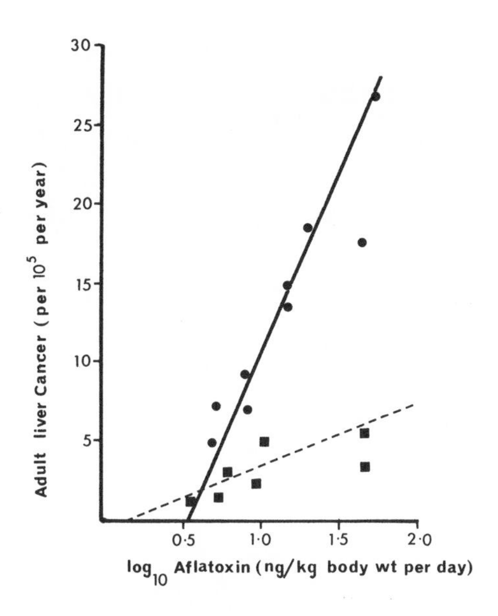

Fig. 5.1. Correlation between the incidence of liver cancer and ingestion of aflatoxin: ■, female; ●, male

Convincing though the epidemiological evidence is, it is necessary to recognize the limitations of such studies. The demonstration of a correlation between two phenomena does not itself prove a causal relationship between them. Infection with the hepatitis B virus is common in those countries with a high incidence of liver carcinoma and it has been shown that a number of viruses are oncogenic. It is, of course, possible that both agents contribute to the occurrence of liver cancer. Diet is a third factor which may influence the activity of aflatoxin, and malnutrition of one form or another is also common in these same regions of the world. The influence of diet may be complex and one form of nutritional deficiency may actually protect a person from the carcinogenic activity of aflatoxin while another form of malnutrition may indeed enhance its activity.

These complex interactions, and the fact that one animal species may be sensitive to aflatoxin and another not, can be rationalized in part by an understanding of what happens to aflatoxin in the animal body. Aflatoxin B_1 is metabolized in a number of different ways to give compounds which may have reduced toxicity, enhanced acute toxicity or which may be carcinogenic (Fig. 5.2). It is probable that aflatoxin B_1 is not itself carcinogenic. Certainly it is not mutagenic in the Ames' test in which a histidine-deficient strain of the bacterium *Salmonella typhimurium* is inoculated onto a histidine-free medium in the presence of the compound to be tested. If the compound is mutagenic then mutations may occur to restore the bacterium's ability to synthesize histidine and it is then able to grow. Although aflatoxin B_1 is not mutagenic a compound formed by incubating it with a preparation of rat liver microsomes, and a system for generating NADPH, is mutagenic and probably carcinogenic. Although it has not been characterized, the compound formed in the liver of rats is almost certainly the epoxide shown in Fig. 5.2 which is thought to react readily with the guanine residues in DNA. Although the epoxide itself has not been isolated the guanine aduct shown in Fig. 5.3 has been purified by hydrolysis of the DNA extracted from an animal treated with aflatoxin B_1.

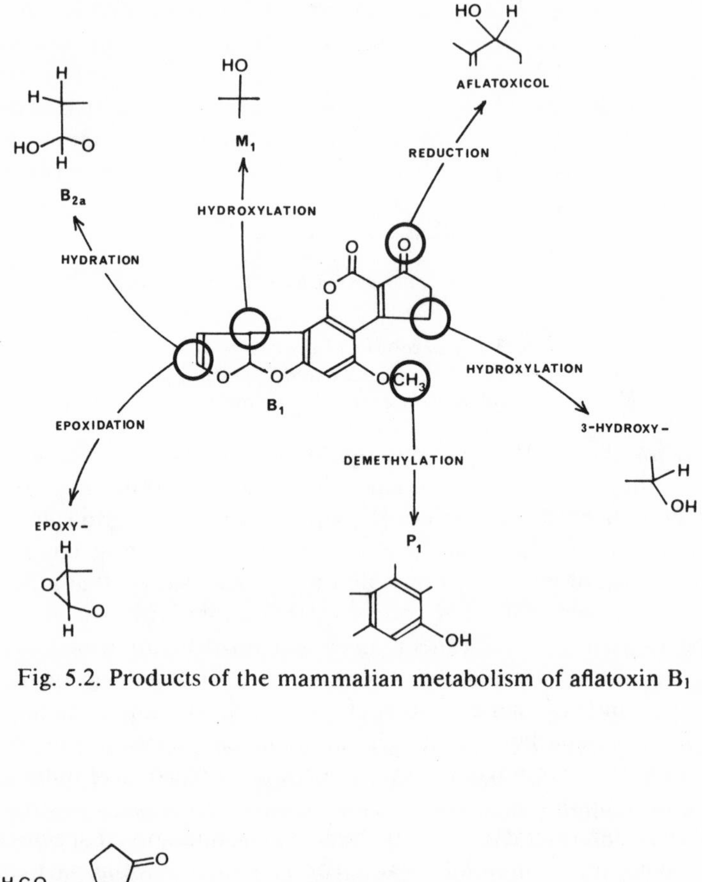

Fig. 5.2. Products of the mammalian metabolism of aflatoxin B_1

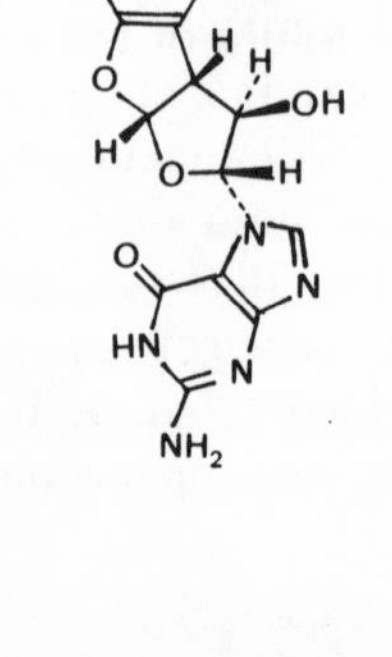

Fig. 5.3. Structure of the aflatoxin-guanine adduct

It is also probable that aflatoxin B_1 itself it not the molecule responsible for the acute damage observed when it is consumed. The hemiacetal derivative known as aflatoxin B_{2a} is very reactive towards proteins being able to bind to two free amino groups (Fig. 5.4). This compound is readily prepared in the laboratory and, when fed to a sensitive animal, appears to be without significant toxicity compared with aflatoxin B_1 itself. But this is almost certainly because it is so reactive that it is effectively neutralized before it reaches any sensitive animal tissue. If, however, it is formed *in situ*, then it can damage the cells in which it is produced by reacting with key enzymes. Certainly it is those animals which have an active system within their liver cells capable of generating aflatoxin B_{2a} which are most sensitive to the acute toxicity of aflatoxin B_1. It should be possible to know what happens in man's liver and, then, confirm whether aflatoxin B_1 can indeed be a causal agent of liver cancer.

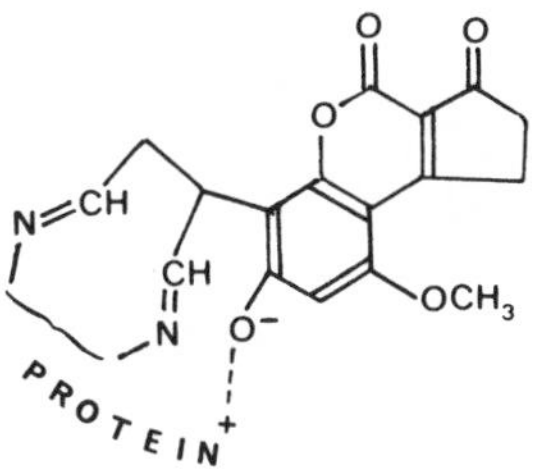

Fig. 5.4. Structure of a hypothetical adduct between a protein and aflatoxin B_{2a}

AFLATOXIN AND OTHER DISEASES

Important though the carcinogenicity of the aflatoxins is, it is necessary to remember that they are also acute poisons. It was, of course, the outbreak of acute liver disease with a very high death rate in turkeys which alerted man to the importance of aflatoxin as an environmental hazard.

At the end of 1974 a large number of villages in northwest India suffered outbreaks of epidemic jaundice involving severe liver disease and the deaths of more than one hundred people. Two independent studies of different regions demonstrated that the population had been exposed to high doses of aflatoxin and that these people had suffered from acute aflatoxicosis. Several other incidents of death associated with exposure to aflatoxin in food have been reported and they all involve concentrations of the order of 0.5 to 2.0 mg aflatoxin per kg food. The food was inevitably a plant product, forming a major part of the diet, such as groundnut meal, maize, rice or cassava.

On 4 November 1968, a three-year old boy was admitted to hospital in Northeast Thailand suffering from fever, vomiting, coma and convulsions. He died six hours later, and autopsy revealed oedema in the brain with degradation of neurones as well as changes in the fatty tissue associated with the liver, kidneys and heart. These symptoms are very similar to those associated with a childhood illness of unknown aetiology, occurring in Australia and known as Reye's syndrome, which also had a high mortality. The death of the child in Thailand was immediately investigated and there seemed to be no doubt that he had consumed a large amount of aflatoxin in heavily contaminated boiled rice. Several other instances of Reye's syndrome in other parts of the world have been studied in an attempt to understand its cause and, although the involvement of aflatoxin in some cases cannot be excluded, it is almost certainly not the only causal agent.

EXPOSURE TO AFLATOXIN IN MAN

Almost all of the documented cases of both chronic and acute aflatoxicosis involve the presence of aflatoxin in food plant products following infection with either *Aspergillus flavus* or *A. parasiticus* in the field, during harvest or during storage. Aflatoxin M_1, secreted in milk by cows consuming feed contaminated with aflatoxin B_1, has been detected in dairy products during surveys in several countries, including the United Kingdom (Anon, 1980). The efficiency of conversion and secretion is not very high so that animals receiving a daily intake of about 20 mg aflatoxin B_1 in their feed secrete milk containing about 10 $\mu g\ l^{-1}$ of aflatoxin M_1. Nevertheless, studies of cirrhosis of the liver in a rural area of Iran implicated aflatoxin M_1 present in cows milk in exceptionally high concentrations (50–500 $\mu g\ l^{-1}$). There are, unfortunately, no records of the levels of contamination of the animal feeds in this instance but they must have been incredibly high. Although aflatoxin M_1 is almost as acutely toxic as B_1, it is less than one hundredth as carcinogenic to rats. It has been estimated that the average per capita daily intake of aflatoxin B_1 in the USA is about 20 ng kg^{-1} body weight (mainly from maize and maize products), whereas the average per capita intake of M_1 is about 0.8 ng kg^{-1} body weight. So, whatever the absolute risk factors are, the relative risk factors for liver cancer from B_1 to M_1 must be of the order of 600 to 1.

Two special groups of humans may be involved with an occupational exposure to aflatoxins unless special precautions are taken: those who handle bulk plant commodities such as cereals, groundnut meal and animal feeds; and those involved in laboratory studies and the analysis of aflatoxins. There has been some evidence of an increased incidence of cancer in workers exposed for two to nine years to dust from an oilseed crushing mill which handled groundnuts. Analyses of the dust in such an environment have provided figures of 0.87 to 72 ng aflatoxin per cubic metre of air with a corresponding respiratory exposure of 39 ng to 3.2 μg per worker per week. Insufficient is known about the toxicology of aflatoxin in man to provide risk factors for such exposure but the limited epidemiological studies suggest that they are significant.

THE GENUS *FUSARIUM* AND DISEASE OF MAN

Outbreaks of the dreadful disease known as alimentary toxic aleukia (but also referred to as septic angina, acute myelotoxicosis and by several other names), which occurred during famine conditions in parts of the Soviet Union during 1942–1947, have been extensively reviewed (see, for example, Joffe, 1983). The disease has actually been recorded in Russia from time to time since the nineteenth century.

Consumption of badly moulded cereals, now considered to have been contaminated with trichothecenes, first causes a discomfort of the mouth, throat and stomach followed by inflammation of the intestinal mucosa. Many symptoms associated with damage to the mucosal membrane systems occur including vomiting and diarrhoea. As more toxin is consumed, damage to the bone marrow and haematopoietic system (responsible for producing blood cells) occurs followed by anaemia and a drop in erythrocyte and platelet counts. Further inflammatory degeneration develops, blood capillary walls become weakened and haemorrhaging frequent. Those areas originally affected become necrotic and the dead tissue becomes infected with bacteria. The trichothecenes, like several other mycotoxins, are immunosuppressive so the body's defences against general bacterial and viral infection are weakened. Many deaths occurred from these secondary infections but continued ingestion of the toxic food would, inevitably, lead to death anyway.

Fortunately, improved harvesting and storage would seem to have eliminated this disease, at least from the regions of the Soviet Union where it had such a devastating impact. Poisoning from *Fusarium* toxins or, indeed, trichothecenes from other mould genera, may still occur in other parts of the world. Thus, the disease associated with 'yellow rain', a subject which has unfortunately become misted by the veils of super power politics, may be another instance of disease initiated by *Fusarium* toxins. (Readers are referred to leading articles in *Science* and *Nature* for some insight into this intriguing phenomenon, e.g. Wade, 1981; Maddox, 1984; Nowicke and Meselson, 1984.)

In Japan, there are many records of nausea, vomiting and diarrhoea associated with the contamination of wheat and rice by species of *Fusarium*. They have frequently been referred to as red-mould disease. Whereas alimentary toxic aleukia probably involved *Fusarium sporotrichioides* and related species, producing primarily T-2 toxin, red mould disease involved a different range of *Fusarium* species the most predominant being *F. graminearum* producing nivalenol, deoxynivalenol and related trichothecenes.

Because of the difficulties of routinely analysing trichothecenes, results indicating possible levels of exposure are not readily available. Table 5.3 provides some figures based on results presented in a series of chapters included in a recent textbook on trichothecenes (Ueno, 1983).

In the Transkei in Southern Africa, North China and Iran there are areas with a high incidence of oesophageal cancer in man. It has been suggested

Table 5.3. Occurrence of trichothecenes in different parts of the world

Country	Date	No. of samples	Commodity	% Cont.*	Toxin and mean conc. (ppm)
USA	1977	52	Maize	46	DON (5.0)
Japan	1970–80	130	Barley/wheat	81	DON (2.5) NIV (1.5)
Canada	1980	45	Wheat	98	DON (0.01–4.3)
Finland	1972	160	Cereals	2	T-2 (0.01–0.03)
Hungary	1976–79	464	Animal feeds	6	T-2 (0.5–5)
Denmark	1980	36	Cereals	2.8	DON (1.0)

DON = deoxynivalenol, NIV = nivalenol. * % of samples contaminated.

that yet another species of *Fusarium*, *F. moniliforme*, may be associated with this condition and, recently, a mutagenic compound known as fusarin C has been isolated from strains of this species and chemically characterized (Fig. 5.5) and it is interesting to note that this compound is an epoxide.

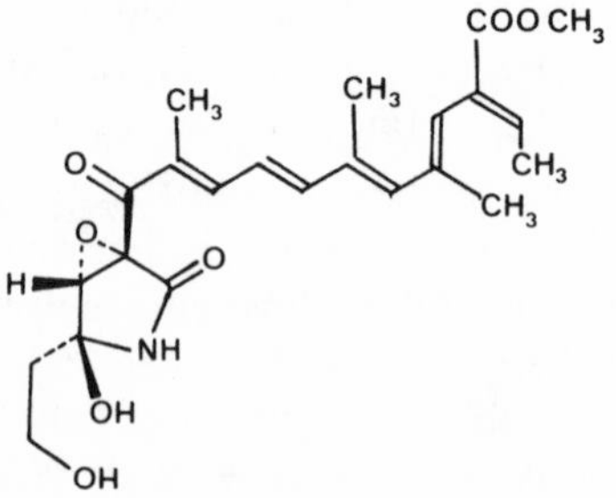

Fig. 5.5. Structure of fusarin C, a mutagenic metabolite of *Fusarium moniliforme*

In China, Eastern USSR and North Korea there is an endemic osteoarthritic disease of man known as Kaschin-Beck's disease which seems to be associated with maize and wheat and may involve infection with strains of *Fusarium*. There is no information about the nature of any toxic metabolites which may be involved but it is interesting that *Fusarium*-infected feeds have also been associated with a bone abnormality in chicks known as tibial dyschondroplasia already described in Chapter 4 (Mirocha, 1984).

Although zearalenone, a metabolite of several species of *Fusarium*, is known to cause a rather specific, and sometimes serious, oestrogenic syndrome in pigs, as well as infertility in cattle, there have not been any reports of adverse effects in man following the consumption of naturally infected food. Zearalenone has been detected in maize meal and cornflakes used for human consumption so it is essential to continue to monitor its presence in the human diet and assess the long-term effects of exposure.

PENICILLIUM TOXINS AND HUMAN DISEASE

That rice coloured yellow by mould activity could be associated with ill health was recognized in Japan during the last century. Specifically, there was an early awareness that moulds may be responsible for cardiac beri-beri but it was not until 1938 that *Penicillium citreo-viride* (= *P. toxicarium*) was shown to be the most likely cause. Other species of pigmented penicillia growing on rice have been isolated and associated with particular disease syndromes recognized in Japan (Table 5.4).

Table 5.4. Species of *Penicillium* and their metabolites associated with different aspects of 'yellow rice' toxicoses

Species	Toxins	Disease
P. citreo-viride	Citreoviridin	Cardiac beri-beri
P. islandicum	Luteoskyrin Cyclochlorotine	Hepatotoxicity
P. citrinum	Citrinin	Nephrotoxicity

Members of the genus *Penicillium* are widely used in a number of countries in the mould ripening of cheeses and meat products, such as sausages. These small greenish-blue moulds are familiar in the moulding of foods with a reduced water activity such as bread and fruit preserves. It is possible to argue that man has been exposed to such moulds and their products for centuries without undue hazard but a continuous awareness of their potential danger is essential. Problems in the occurrence of mycotoxins in some food commodities may arise from changes in food and agricultural technology. In the latter half of the twentieth century these changes are occurring with increasing rapidity and on an increasingly large scale making it essential to maintain a continuous awareness of the potential that moulds have for producing toxic metabolites in our environment.

FURTHER READING

Anon (1979). Mycotoxins, *Environmental Health Criteria*, **11**, pp. 96-98, World Health Organisation, Geneva.

Anon (1980). *Survey of Mycotoxins in the United Kingdom*, Fourth Report of the Steering Group in Food Surveillance, HMSO, London.

Austwick, P. K. C. (975). Balkan nephropathy. *Proceedings of the Royal Society of Medicine*, **68**, 219–221.

Austwick, P. K. C. (1984). Human mycotoxicosis—past, present and future. *Chemistry and Industry*, 547–551.

Barnes, J. M. (1967). *Foundation Study Group No. 30*, Ciba Foundation, London, p. 111.

Joffe, A. Z. (1983). Food borne diseases: Alimentary toxic aleukia. In: *Handbook of Foodborne Disease of Biological Origin*, M. Rechcigle (Ed.), CRC Press, Florida.

Krogh, P., Hald, B., Plestina, R. and Ceovic, S. (1977). Balkan (endemic) nephropathy and foodborne ochratoxin: Preliminary results of a survey of foodstuffs. *Acta path. microbiol. Scand. Sect. B*, **85**, 238–240.

Le Breton, E., Frayssinet, C. and Boy, J. (1962). Sur l'apparition d'hetomes 'spontanes' chez le Rat Wistar. Role de la toxine de l'*Aspergillus flavus.* Interet en pathologie humaine et cancerologie experimentalle. *Compte rendu Academie Science (Paris)*, **225**, 784–786.

Maddox, J. (1984). Natural history of yellow rain. *Nature, (London)* **309**, 207.

Mirocha, C. J. (1984). Mycotoxicoses associated with *Fusarium.* In: *The Applied Mycology of Fusarium*, M. O. Moss and J. E. Smith (Eds), Cambridge University Press, Cambridge, pp. 141–155.

Newberne, P. M. and Wogan, G. N. (1968). Sequential morphological changes in aflatoxin B_1 carcinogenesis in the rat. *Cancer Research*, **28**, 770–781.

Nowicke, J. W. and Meselson, M. (1984). Yellow rain—a palynological analysis. *Nature (London)*, **309**, 205–206.

Oettle, A. G. (1965). The aetiology of liver carcinoma in Africa with an outline of the mycotoxin hypothesis. *South African Medical Journal*, **39**, 817–825.

Peers, F. G. and Linsell, C. A. (1977). Dietary aflatoxins and human primary liver cancer. *Ann. Nutr. Aliment.*, **31**, 1005–1018.

Stoloff, L. (1977). Aflatoxins—an overview. In: *Mycotoxins in Human and Animal Health*, J. V. Rodricks, C. W. Hesseltine and M. A. Mehlmann (Eds), Pathotox, Illinois, pp. 7–29

Ueno, Y. (Ed.) (1983). *Trichothecenes, Chemical, Biological and Toxicological Aspects*, Elsevier, Amsterdam.

Wade, N. (1981). Yellow rain and the cloud of chemical war. *Science*, **214**, 1008–1009.

CHAPTER 6

Natural Occurrence of Mycotoxins

Mycotoxins can enter the human and animal dietary systems by indirect or direct contamination. Indirect contamination of foods or feeds can occur when an ingredient of a process has previously been contaminated with toxin-producing moulds and, although the moulds may be killed or removed during processing, mycotoxins may often remain in the final product. Such contamination of cereals and oil seeds represents the main point of entry of many mycotoxins into food and feed chains. In direct contamination the product later becomes infected with a toxigenic mould with subsequent toxin formation. Almost all foods and feeds can be susceptible to mould growth at some stage during their production, processing, transport and storage.

Contamination of finished foods and feeds with moulds invariably leads to the exclusion of that product from the food chain in developed economies. However, in most developing countries this is generally not possible and mouldered foods such as cereal products are often a regular part of the daily diet. In parts of Africa where liver cancer is extremely prevalent regular occurrence of aflatoxins and *Fusarium* toxins in the cereal foods has been recognized. Undoubtedly stricter adherence to good food and feed hygiene would greatly reduce mycotoxin presence. Mycotoxin residues in animal tissues, e.g. kidneys, and animal products, e.g. milk, can occur by consumption of mycotoxin-contaminated feeds by the animals.

In practice, all ingredients destined for human or animal feeding will have been exposed at some point in time to mould contamination. The nature and extent of toxigenic mould contamination will determine the presence or absence of mycotoxins in the product. While the identification of contaminating fungi can be of diagnostic value in outbreaks of mycotoxicosis, positive conclusions can only be deduced by extraction and identification of the suspected toxin(s), since:

1. The presence of the fungus is no assurance that it was producing the toxin.
2. A given toxin may persist in a product when the fungus is no longer present.
3. A given fungus may be capable of producing more than one toxin.
4. A given toxin may be produced by different genera of fungi (see Chapter 2).

However, the presence of a toxigenic fungus undoubtedly gives some pointer to a potential hazard. Table 6.1 outlines the main possible routes for mycotoxin entry into human and animal foods.

Table 6.1. Possible routes for mycotoxin contamination of human and animal foods (adapted from Jarvis, 1976)

Mould damaged foodstuffs
(a) Agricultural products, e.g.
cereals
oilseeds (groundnuts)
fruits
vegetables
(b) Consumer foods (secondary infections)
Compounded animal feeds (secondary infections)
Residues in animal tissues and animal products, e.g.
milk
dairy produce
meat
Mould-ripened foods, e.g.
cheeses
fermented meat products
oriental fermentations
Fermented products, e.g.
microbial proteins
enzymes
food additives such as vitamins

AGRICULTURAL CROPS

Nuts and oilseeds

Nuts and oilseeds are the most thoroughly studied of all commodities associated with mycotoxins, especially aflatoxin. Because of its high protein content the groundnut has become a major component of many animal diets and consequently groundnut production has become a major agricultural industry in many warm and humid countries supplying not only their own needs but catering also for the expanding animal feed industry particularly in developed nations. Contamination of peanuts occurs primarily in the field during harvest when the nuts are being dried. Mechanical damage, insect damage and excessive rain during the drying period are all contributory factors to invasion by *A. flavus*. The statistics of aflatoxin presence in groundnuts are uniformly incriminating with surveys carried out throughout the world regularly indicating disturbing levels of aflatoxin.

The incidence and levels reported vary markedly from one geographical area to another (Table 6.2). In a recent survey the most highly contaminated samples originated from India, Gambia and Malawi and to a lesser extent from Brazil and Egypt. Even samples from the USA can be appreciably contaminated. Serious contamination in Nigerian and Australian peanuts has been a recurring problem when harvesting is associated with wet weather. Many countries now restrict or ban the importation of groundnut meal for animal feeds. Peanut butter may also show variable levels of aflatoxin contamination.

With other types of nuts only Brazil nuts have shown regular occurrence of aflatoxins; almonds, cashew nuts, hazelnuts and walnuts seem to be singularly free from contamination, most probably due to shell texture and methods of harvesting. Users of confectionary-grade nuts have long been aware of the problems associated with aflatoxin contamination and manufacturers rigorously use testing and selection procedures at all stages from import of nuts to final product assessment.

Raw and roasted cocoa beans and presscake can contain low levels of aflatoxins and ochratoxin A but, to date, no mycotoxin has been detected in chocolate. Ochratoxin A has been detected in low concentrations in green coffee beans but was largely destroyed during the roasting process (Table 6.3). No mycotoxins have been detected in commercial coffee products.

Aflatoxin contamination of cottonseeds is a major worldwide problem and levels as high as 200 000 to 300 000 $\mu g\ kg^{-1}$ have been demonstrated in individual lots. Most of the contamination occurs in the field and is caused by a combination of high ambient temperature, high humidity and mechanical damage to the bolls before and during drying. When contaminated cottonseed is processed for oil most of the aflatoxin is concentrated in the residual meal and should this meal be used as an ingredient of animal feeds serious toxicity problems can develop. Human exposure to aflatoxins can arise in countries using unrefined oil from peanuts and cottonseed for human consumption.

Aflatoxin contamination of soya and soya products is rare in commerce in the USA but detectable levels have been demonstrated in edible beans in Africa and Thailand. Significant levels of contamination with ochratoxin A have been observed in soya beans and raw soy flour. Ochratoxin A contamination of fermented soya beans, e.g. soya sauce, miso, etc., has never been documented.

Table 6.2. Aflatoxin content of peanuts as imported into the UK during a single year (1977/78 season)[a] (Anon, 1980)

Country of origin	No. of samples[f] analysed	Number of samples containing levels of total aflatoxin[b] in the stated range ($\mu g\ kg^{-1}$)								
		0[c]	1–10	11–30	31–50	51–100	101–200	201–300	301–400	400+
Brazil	2	0	0	0	2	0	0	0	0	0
Egypt	2	1	0	0	0	1	0	0	0	0
Gambia	5	0	2	1	1	0	0	0	0	1[d]
India	35	18	2	2	2	6	3	1	0	1[e]
Malawi	53	27	9	8	2	5	1	1	0	0
S. Africa	6	6	0	0	0	0	0	0	0	0
USA	56	43	7	2	3	0	1	0	0	0
Total	159	95	20	13	0	12	5	2	0	2

[a] Any variation which may occur from year to year is under investigation.
[b] In most cases aflatoxin B_1 was the major contaminant.
[c] Not detected at 0.5 $\mu g\ kg^{-1}$.
[d] The maximum level measured was 440 $\mu g\ kg^{-1}$.
[e] The maximum level measured was 410 $\mu g\ kg^{-1}$.
[f] 20 kg samples taken from each 20 tonne batch.

Fruit products

The occurrence of patulin in moulded fruit has been widely recognized with levels up to several hundred $\mu g\ kg^{-1}$ in some pressed apple juices. Apple juices and grapefruit juices can frequently be contaminated with patulin. Orange and other juices, squashes, pickles, sauces and other fruit products have seldom shown contamination with patulin. Patulin can also occur in inadequately processed silage.

Cereals

Cereals can support fungal growth while still in the field, during storage and after processing into food products or animal feed. Small grains (sorghum, barley, oats, wheat, rye and rice) unless badly handled in storage or after preparation appear to be less susceptible to mycotoxin formation than are larger grains such as maize (also known as corn in the USA). It is well documented that mycotoxin contamination is more regularly associated with low-grade cereals which normally do not enter the human food chain in developed countries. Regrettably this is not true in many developing countries where the better quality cereals are often shipped abroad. The main mycotoxins that have been detected in cereals include aflatoxin, sterigmatocystin, ochratoxin A, zearalenone, T-2 toxin and vomitoxin. The incidence of occurrence varies with climatic conditions prevailing at the time of harvest and during transportation and later storage.

Many mycotoxins are formed in cereals as the plant grows and matures. Aflatoxin production was originally considered to be mainly associated with stored agricultural crops but it is now apparent that *A. flavus* can occur on many crops, e.g. peanuts, maize, cottonseed, before and during harvest. Although insect damage is one of the main factors allowing aflatoxin-producing fungi to grow into plant tissues many other promoting conditions have now been identified, e.g. stress factors on the growing plant, mechanical damage, mineral nutrition deficiencies and unseasonal temperatures.

Aflatoxin is regularly detected in maize throughout the world and recent serious contamination in USA (1983) was associated with drought conditions and subsequent insect infestation of the growing crop. The regularity of aflatoxin formation in dent corn over a period of five years is shown in Table 6.4. Analysis of maize samples from Thailand, the Philippines and South Africa regularly show disturbingly high levels of contamination (up to 500 $\mu g\ kg^{-1}$). In the USA over 80 per cent of the human exposure to aflatoxin in the diet comes from the consumption of maize and maize products. Ochratoxin A has been detected in maize and maize products. Ochratoxin A has been detected in maize, wheat, oats, rye and particularly barley in many parts of the world (Table 6.3).

Fusarium fungi with associated mycotoxins, zearalenone and the trichothecenes predominantly occur during the growing phases of many

Table 6.3. Natural occurrence of ochratoxin A in foods and feeds of plant origin (Krogh and Nesheim, 1982)

Commodity	Country	Number of samples	Per cent contaminated	Ochratoxin A level (range $\mu g\ kg^{-1}$)
Foods				
Maize	USA	293	1.0	83–166
Maize (1973)	France	463	2.6	15–200
Maize (1974)	France	461	1.3	20–200
Wheat (red winter)	USA	291	1.0	5–115
Wheat (red spring)	USA	286	2.8	5–115
Barley (malt)	Denmark [d]	50	6.0	9–189
Barley	USA	182	12.6	10–29
Coffee beans	USA	267	7.1	20–360
Maize	Yugoslavia [a]	542	8.3 [b]	6–140
Wheat	Yugoslavia [a]	130	8.5 [b]	14–135
Wheat bread	Yugoslavia [a]	32	18.8 [b]	
Barley	Yugoslavia [a]	64	12.5 [b]	14–27
Barley	Czechoslovakia	48	2.1	3800
Bread	UK [c]	50	2	710
Flour	UK	7	28.5	490–2900
Rice	Japan [c]	2	(100)	230–430
Beans	Sweden	71	8.5	10–442
Peas	Sweden	72	2.8	10
Feeds				
Barley, wheat, oats, rye, maize	Poland	150	5.3	50–200
Mixed feed	Poland	203	4.9	10–50
Maize	Yugoslavia	191	25.7	45–5125
Barley, oats	Sweden	84	8.3	16–409
Wheat, hay	Canada [c]	95	7.4	30–6000
Wheat, oats, barley, rye	Canada [c]	32	56.3	30–27000
Barley, oats	Denmark [c]	33	57.6	28–27500

[a] From an area with endemic human nephropathy.
[b] Average values for a period of 2–5 years.
[c] All samples suspected of containing mycotoxins.
[d] Unpublished data.

Table 6.4. Aflatoxin levels in dent corn grown in Virginia, 1976–1980 (Shotwell and Hesseltine, 1983)

	Collected from trucks										Collected at harvest			
	1976		1977		1978		1979		1980		1978		1979	
Total aflatoxin ng g^{-1}	No. of samples	(%)	No. of samples	(%)	No. of samples	(%)	No. of samples	(%)	No. of samples	(%)	No. of samples	(%)	No. of samples	(%)
ND	77	(63)	52	(51)	63	(64)	81	(71)	18	(18)	79	(88)	93	(79)
<20	13	(10)	17	(17)	10	(10)	13	(11)	20	(20)	2	(2)	9	(8)
20–100	21	(17)	18	(18)	21	(21)	8	(7)	32	(32)	5	(6)	7	(6)
101–500	9	(7)	10	(10)	5	(5)	10	(9)	26	(26)	4	(4)	7	(6)
501–1000	1	(1)	1	(1)	—	—	2	(2)	1	(1)	—	—	—	—
>1000	2	(2)	3	(3)	—	—	—	—	2	(2)	—	—	1	(1)
Total	123		101		99		114		99		90		117	
% incidence	37		49		36		29		82		12		21	
% ≥20 ng g^{-1}	27		32		26		18		61		10		13	
% >100 ng g^{-1}	10		14		5		11		29		4		7	
Av. level (ng g^{-1}) all samples	48		91		21		34		137		13		36	
Av. level (ng g^{-1}) pos. samples	130		187		58		118		167		110		176	

ND = not detected.

Table 6.5. Natural occurrence of zearalenone (Shotwell, 1977)

Commodity or product	Examined because of	Country	Zearalenone (ppm)
Hay	Infertility in dairy cattle	England	14.0
Feed	Infertility in cattle and swine	Finland	25.0
Corn	Hyperoestrogenism in farm animals	France	2.3
Animal feed	Hyperoestrogenism in cattle and swine	USA	0.1–2900
Corn		Yugoslavia	18
Corn	Poisoning in swine	Yugoslavia	2.5–35.6
Corn	Severe mould damage and swine refusal	Yugoslavia	0.7–14.5
Barley	Stillbirths, neonatal mortality and small litters in swine	Scotland	0.5–0.75
Corn (freshly harvested)	*Gibberella zeae* damage	USA	0.1–1.5
Corn (stored)	*A. flavus* damage	USA	ND-92
Barley	Death in swine	Scotland	'Traces'
Feed	Field problems in animals	USA	Not stated
Grain sorghum	Head blight in sorghum	USA	Not stated
Corn	Swine hyperoestrogenism	Yugoslavia	35.6
Pig feed	Swine hyperoestrogenism	USA	50.0
Corn	Swine hyperoestrogenism	USA	2.7
Sorghum	Cattle abortion	USA	12.0
Corn	Swine abortion	USA	32.0
Silage		USA	87.3
Corn		England	306.0
Corn	Swine feed refusal	USA	2.5
Dairy ration	Cattle feed refusal, lethargy, anaemia	USA	1.0
Pig feed	Swine internal haemorrhaging	USA	0.1
Pig feed	Swine hyperoestrogenism	Yugoslavia	0.5
Pig feed	Swine infertility and abortion	USA	0.01

ND: not detected.

cereals. Their continued development is less likely after harvest and during storage. Zearalenone has been detected in maize, sorghum and barley throughout the world (Table 6.5). The causal *Fusarium* spp. attack and parasitize seeds during periods of heavy rainfall and will proliferate on mature grains that have not dried because of wet weather at harvest or on grains that have been stored wet. Normally low temperatures (12–14° C) are required to initiate and maintain high production of zearalenone. The presence of zearalenone in cereal grains has mainly been established as the result of investigating field outbreaks of mycotoxicosis or from screenings of grain collected at dispersal points in the marketing systems. There is growing evidence that if zearalenone is detected in maize and other cereals there is a high probability that other *Fusarium* toxins may well be present.

Until recently evidence for contamination of cereals by trichothecenes has been extremely limited due to the fact that adequate analytical methodology has only now become widely applied. Four trichothecenes have been well documented in cereals, viz. T-2 toxin, diacetoxyscirpenol, nivalenol and deoxynivalenol (vomitoxin). Their occurrence has been largely in wet temperate climates such as Japan, Northern Europe and parts of North America, particularly Canada (Table 6.6). In Canada, vomitoxin is mainly a problem in Ontario, Quebec and the Atlantic Provinces where warm humid weather encourages the growth of *Fusarium*. The 1982 Ontario wheat crop showed levels of up to 7.0 ppm. Vomitoxin has frequently been associated with maize refused by swine in the USA and vomiting in dogs and cows in South Africa. Detailed studies in South Africa with maize have indicated a very high rate of contamination by *Fusarium* species and the toxins they produce. Samples of maize (generally mouldy) intended for animal feeding or beer brewing showed a much higher rate of contamination together with higher concentrations of toxins than samples intended for human consumption (relatively mould-free).

Table 6.6. Contamination of cereals with trichothecenes (Osborne, 1982)

Commodity	Country	Toxin	Incidence	Level (ppm)
Corn	USA	T-2	1/1	2.0
Corn	USA	T-2?	94/173	0.05–1.0
Mixed feed	USA	DAS, T-2, vomitoxin	9/200	0.04–1.8
Corn	USA	Vomitoxin	24/52	0.5–10
Grain	Finland	T-2	9/230	0.01–0.05
Corn	Austria	Vomitoxin	3/3	1.3–7.9
Corn	Canada	Vomitoxin		
Wheat/barley	Japan	Vomitoxin, nivalenol	?	62.3 71.5
Barley	Japan	Vomitoxin	?	4.2
Corn	S. Africa	Vomitoxin (+F-2)*	?	0.25–7.4
Wheat	UK	Vomitoxin	1.35	0.09

* F-2 = zearalenone

Several other mycotoxins can occur in cereals and forage and consumption by grazing animals can cause serious and even disastrous effects. Sporidesmins are mycotoxins produced by *Pithomyces chartarum* growing on pasture grasses particularly in New Zealand, Australia and some parts of USA. Facial eczema and oedema of the grazing animals occurs on exposure of infected animals to light. Stachybotryotoxicosis occurs primarily in Eastern Europe causing severe haemorrhaging and necrosis in horses consuming hay infected with *Stachybotrys chartarum*. Ergotism is one of the earliest and best documented mycotoxicoses of man and animal. Historically, infection of rye and other cereals by *Claviceps purpurea* resulted in toxic alkaloids passing into the flour that was subsequently used for bread making. Ergotism in humans is now extremely rare but it is still a common problem with grazing animals consuming wild grasses contaminated with *Claviceps*. Infections may also occur in commercial rye crops. Tremorgens produced by various species of *Penicillium* have been isolated from a number of cereals and pomaceous fruits.

Possibly the greatest potential for toxigenic mould growth and mycotoxin production will occur with storage of inadequately dried agricultural products and the rewetting of dried and stored products. Such rewetting can easily occur by water leaking into bins, flooding and more commonly by condensation (see Chapter 8). Cereal grains are often mixed or blended to reduce the overall moisture level. Although a measuring meter will record the overall moisture level of many grains a fungal spore will react only to the microenvironment around a single seed. Storability of cereals and animal feeds depends not only on the average moisture levels in a storage bin but also on the moisture concentration in the individual seeds. With inadequate mixing a few grains with moisture levels able to support fungal growth may generate areas of active growth or 'hot spots' in grain piles. Mycotoxin production may then occur in small localized areas in an otherwise mycotoxin-free storage bin. Subsequent removal of the stored material will give irregular distribution of the mycotoxin(s) with ensuing mycotoxicosis. Under these conditions it is rare for the whole batch to be contaminated and the unequal toxin distribution makes analytical studies difficult.

Duration of storage is an important factor when considering mycotoxin formation. The longer the retention in a storage bin the greater will be the possibility of building up environmental conditions conducive to toxigenic mould proliferation. Bin hygiene seldom rates high in farm management and months, even years, may pass with no routine clean-out. On many farms fresh feed components are often added to old material present in bins—often last year's crop. This probably represents the most serious input of mycotoxins into animal diets and yet, within the agricultural community, it is the least recognized.

Storage of ground feed materials, e.g. wheat, maize, etc., prior to compounding creates special problems. Here the protective outer testa of the

Table 6.7. Mycotoxins in breakfast cereals (Norton *et al.*, 1982)

Type of breakfast cereal	Number of samples tested	Number of samples in which aflatoxins were detected			Number of samples in which ochratoxin A was detected			Number of samples in which zearalenone was detected	
		1–5 $\mu g\ kg^{-1}$	6–10 $\mu g\ kg^{-1}$	11–25 $\mu g\ kg^{-1}$	<10 $\mu g\ kg^{-1}$	11–20 $\mu g\ kg^{-1}$	21–50 $\mu g\ kg^{-1}$	<50 $\mu g\ kg^{-1}$	51–150 $\mu g\ kg^{-1}$
Corn based	6	1	0	0	2	2	0	2	0
Wheat based	14	1	0	0	4	4	1	3{7[a]]	
Oat based	6	1	0	0	2	0	0	4[a]	0
Bran based	6	2	0	0	3	0	0	0	0
Muesli-type ingredients	7	2	0	0	3	0	0	0	0
Rice based	5	3	0	0	1	0	0	0{1[a]]	0
Totals	56	13	0	1	13	8	1	7[b]	1
Percentage of total samples analysed		23	0	2	23	14	2	13	2

[a] Toxin tentatively identified but too much interference for it to be confirmed and precisely quantified.
[b] Excluding those positive but unconfirmed.
All samples analysed by HPLC.
Limits of detection : approximately 7 $\mu g\ kg^{-1}$ for each of the aflatoxins; 10 μg ochratoxin A kg^{-1}; 50 μg zearalenone kg^{-1}.

seeds have been destroyed and the rich nutrients inside can now be easily colonized by toxigenic moulds. Thus, mycotoxins in animal feeds will usually arise from using contaminated raw materials, e.g. aflatoxin-containing groundnut or cottonseed meal, contaminated cereals or by secondary infection by toxigenic moulds resulting from inadequate handling during and after manufacture.

Since normally only high quality cereals are used for human foods it can be expected that levels of mycotoxins entering the human food chain will be significantly less than for animal feeds. Maize products still constitute the main source of mycotoxin entry into human diets, in particular, aflatoxin, ochratoxin A and zearalenone. However, the various processing procedures applied to maize before incorporating into food products drastically reduces the levels of most mycotoxins. A recent UK survey showed that 5 per cent of breakfast cereals analysed contained detectable levels of aflatoxin (less than $5\,\mu g\ kg^{-1}$), 39 per cent contained ochratoxin A (less than $20\,\mu g\ kg^{-1}$) and 14 per cent contained zearalenone (Table 6.7). However, in almost all cases the levels of toxins were predominantly near the lower limits of detection for these toxins.

The current good standards of crop handling and storage in most developed countries does mean that mycotoxin levels in cereal raw materials and products destined for human consumption is usually at the lower levels of detection limits. However, there are isolated examples when much higher, toxin levels can occur. Constant vigilance must always be practised. Warm and humid climatic conditions, coupled with less efficient agricultural and food technology practices, generally ensure much higher levels of mycotoxins entering the human and dietary chains in many developing countries.

ANIMAL PRODUCTS

Milk and Dairy Products

Lactating mammals such as cows, sheep and goats consuming feedstuffs contaminated with aflatoxin B_1 or B_2 excrete into their milk and urine the toxic aflatoxins M_1 or M_2 (the milk toxins). Aflatoxin M_1 is of particular significance being derived from aflatoxin B_1 by hydroxylation at the tertiary carbon of the difuran ring system. Aflatoxin M_1 has also been identified in extracts of moulded peanuts.

The oral 7-day LD_{50} of M_1 in ducks is similar to that of B_1 while the hepatocarcinogenicity is of a lower potency. The transfer ratio (i.e. consumed B_1 to excreted M_1) is approximately 200:1. Aflatoxin M_1 appears in the milk within 12 h after consumption of B_1 by the cow; excretion decreases rapidly about one day after feeding of toxin has stopped while small amounts will still be excreted up to 3–4 days later.

The natural occurrence of aflatoxin M_1 in commercial milk outlets has been documented in many countries (Table 6.8). A seasonal effect has been demonstrated in most surveys with lower levels occurring during summer months when animals were consuming more grass than concentrated feeds. In the UK regulations limit the permissible level of B_1 in dairy feed to 20 $\mu g\ kg^{-1}$ which in theory should give an approximate maximum level of M_1 in milk of less than 0.1 $\mu g\ l^{-1}$ (Anon, 1980). In Europe the main source of B_1 contamination in dairy rations has been shown to be aflatoxin B_1-contaminated groundnut meal. The embargo on groundnut meal for animal feed introduced by the British Government in 1981 has resulted in a dramatic decrease in M_1 levels in milk particularly in the northern areas where cattle spend longer periods on concentrated feeds.

Table 6.8. Incidences of aflatoxin M in milk (Brown, 1982)

Country	No. positive/ no. samples	Proportion positive (%)	Range of concentrations ($\mu g\ L^{-1}$)
South Africa	5/21	23	<0.02–0.2
West Germany	28/61	46	0.04–0.25
West Germany	118/260	45	0.05–0.30
West Germany	79/419	19	0.05–0.54
USA	191/302	63	T–>0.7
Belgium	42/68	62	0.02–0.2
India	3/21	14	up to 13.3
East Germany (winter)	4/36	11	1.7–6.5
East Germany (summer)	0/12	0	–
Netherlands (winter)	74/95	82	0.03–0.5
UK	85/278	31	0.03–0.52

Relatively high levels of M_1 in some American milk samples prompted the Food and Drug Administration to set a maximum allowable M_1 level of 0.5 $\mu g\ l^{-1}$ in fluid milk entering interstate commerce. Aflatoxin B_1-contaminated cottonseed meal and maize are the main sources of toxin entering into the feed concentrates.

The concentration of milk into dried milk products results in a marked increase in the levels of M_1—up to eight-fold enrichment. Measurable levels have been detected in some baby foods and must give cause for considerable concern. Processing of milk can reduce the amount of detectable residual M_1, viz. up to 65 per cent reduction with pasteurization at 72° C for 45 s, 81 per cent when sterilized at 115° C and 86 per cent when spray-dried.

Separation of the components of milk for the preparation of dairy products such as cream, butter, cheese and whey results in a specific distribution of M_1 due to its insolubility in the milk fat and absorption to the curd. In butter production 10 per cent of the M_1 content of milk goes into the cream while most is retained in the skimmed milk. Ten per cent of the M_1 in the cream will ultimately appear in the butter, the remainder staying in the buttermilk. In the production of cottage cheese the preliminary acidification causes about 20 per cent loss of M_1 but of the remainder 30 per cent is concentrated into the curd and 50 per cent in the whey. The wide usage of whey in ice-cream, meat products, etc., will require regular monitoring for the presence of M_1.

Prior to the restriction on the use of groundnut meal in dairy concentrates the presence of aflatoxin M_1 was not uncommon in the UK home-produced cheeses with documented levels of 0.1–0.4 $\mu g\ kg^{-1}$. Many continental cheeses may also on occasion have low levels of aflatoxin M_1. In USA following the introduction of guidelines for milk there has been a near exclusion of the milk toxin in cheese, ice-cream, yoghurt and non-fat milk solids. Clearly the main approach to reducing the occurrence of the milk toxins is by excluding aflatoxin B_1 from animal feeds.

Animal tissues

There is apparently little or no danger in the carry-over of mycotoxins into the muscle tissues of most animals consuming feeds contaminated with mycotoxins since the transfer ratios are obviously high. For aflatoxin B_1, the main mycotoxin analysed, transfer ratios ($\mu g\ kg^{-1}$ mycotoxin in feed: $\mu g\ kg^{-1}$ mycotoxin in tissue) can range between 1000 and 14 000 for muscle tissue. Any animal exposed to such high levels of mycotoxins would have shown obvious disease symptoms or even death and would not have entered the human food chain. Exceptions to these high transfer ratios can sometimes occur and, in particular, pigs suffering from ochratoxin A-induced nephropathy can have measurable levels of ochratoxin in their kidneys. Similarly, aflatoxin B_1 has been found in livers and kidneys of pigs in feeding trials with B_1 contaminated diets. T-2 toxin could not be detected (analytical limit 25 $\mu g\ kg^{-1}$) in the muscle tissues of chickens exhibiting oral lesions and haemorrhages of the intestine and kidneys when consuming feed containing up to 10 or 15 mg T-2 toxin kg^{-1}. In general, therefore, animal tissues do not constitute a natural source of mycotoxin contamination in the human food chain.

MOULD-RIPENED FOODS

Mould fungi have a long history in their use in the production of oriental foods and sauces, e.g. soy sauce, tempeh, miso; in cheeses, e.g. Stilton, Roquefort, Camembert, Danish Blue; and in certain meat products, e.g. continental sausages. Such products are produced both by commercial concerns and by home or cottage industry. The potential for toxigenic moulds to occur in these fermentations has been extensively studied and debated with little evidence being demonstrated by actual presence of the well-known toxic metabolites.

Oriental fermentations

In Japan, China, Taiwan, Korea and other South east Asian countries, *Aspergillus oryzae* and related species have been used for the production of shoyu (soy sauce), miso, sake, muru, amasake and rice vinegar while in Indonesia, tempeh is prepared by growing *Rhizopus* spp. on soybeans or peanuts.

The *Aspergillus* spp. used in many of these fermentations are close relatives to the toxigenic moulds *A. flavus* and *A. parasiticus*. It is because of this taxonomic similarity that much care must be taken to ensure the correct identification of producer strains. However, in Japan, careful studies have failed to demonstrate aflatoxin-producing strains among the *Aspergillus* spp. used by the main manufacturers for food fermentations. Although some cultures produced fluorescent compounds having R_f values resembling those of aflatoxins, further chemical analysis indicated that they were, in fact, several kinds of non-toxic pyrazine compounds including flavacol, isocoumarin compounds, lumichrome and compounds other than aflatoxins.

In Japan a wide range of industrially used *Aspergillus* spp. have been checked for possible production of aflatoxins, sterigmatocystin, ochratoxin A, patulin, penicillic acid and cyclopiazonic acid and with the exception of a few strains producing cyclopiazonic acid all proved negative. Many of these strains were able to produce secondary metabolites such as aspergillic acid, kojic acid, 8-nitropropionic acid and oxalic acid but only at levels which did not constitute any toxic hazard to humans. The *Rhizopus* strains used in tempeh fermentations have no historical record of toxicity.

Cheeses and fermented meat products

Many recognized cheeses owe some part of their distinctive flavour to the involvement of *Penicillium* strains, e.g. *P. roquefortii, P. camembertii.* Although they will produce many secondary metabolites as a consequence of their growth in or on the cheese, the long-term toxicity of these metabolites remains unproven. A similar situation also exists for the *Penicillium* fungi used in certain meat products.

Mould spoilage of dairy products, particularly of cheeses, has been regularly recorded. The predominant contaminants are mainly species of *Penicillium* and *Aspergillus* many of which are potential mycotoxin producers. Surveys have shown infrequent occurrence in low concentrations of several mycotoxins including ochratoxin A, sterigmatocystin, mycophenolic acid, cyclopiazonic acid, patulin, citrinin and penicillic acid. In the primary ripening of Gouda cheeses in warehouses fungal contamination of the outer rind showed significant diffusion of sterigmatocystin into the outer rind. However, the washing processes prior to final coating for commercial sale will normally remove most of the toxin. In practice mould-spoiled cheeses should not be consumed.

Table 6.9. Maximum tolerated aflatoxin levels in foods and feeds in various countries (Schuller *et al.*, 1982)

Country	Commodity	Established tolerance[a] ($ng\ g^{-1}$)	Based on detection limit[b]	Legal basis	Remarks
Australia	All foods	5		Recommendation National Health and Medical Research Council, Oct 1976	
Australia (Victoria)	Peanuts, peanut products	15		Victoria Food and Drug Administration Standards (Amendment 17 Regulation 1977, Statutory Rule 373# 1977)	
Austria	All feeds	50		Futtermittelverordnung 1981 1981 Bundesgesetzblatt 48/1981-17.03.1981	
Belgium	All foods	5(B_1)	+	Koninklijk Besluit - 03.02.1975 Benelux Aanbeveling M(77)5- 03.05.1977	Member of European Communities (EC)
	Milk, milk products	1(M_1)	+	Koninklijk Besluit - 03.01.1975	
	All feeds	see EC		see EC	
Canada	Nuts, nut products	15		Regulation B.01.047(n)	Under general authority of Food and Drug Act
Denmark	Peanuts, peanut products	10		Indenrigsmin. bekendtgørelse- 07.01.1971	Member of European Communities
European Communities	Straight feedstuffs	50(B_1)		EC Directive 74/63/EEC of 17.12.1973 published in : Official Journal of the European Communities L38, 22.02.1974, p.11.	Includes Belgium, Denmark, Federal Republic of Germany, France, Greece, Ireland, Italy, Luxembourg, The Netherlands, United Kingdom
	Complete feedingstuffs not for dairy cattle, calves, lambs	50(B_1)			
	Complete feedingstuffs for pigs and poultry	20(B_1)			
	Other complete feedingstuffs	10(B_1)			
	Complementary feedingstuffs for dairy cattle	20(B_1)			
GDR	Peanuts, peanut products	10 or 5 (B_1)		Aflatoxinverordnung Bundesgesundheitsblatt (BGB1) IS 3313-30.11	Member of European Communities

India	Peanut meal (food) Peanut meal (feed for export)	30 1,000			P.Krogh, *Pure Appl. Chem.*, **49**, 1977, 1719.
Japan	All foods	10(B_1)		Food Sanitation Investigation Council April 1974	Not clear whether B_1 or $B_1 + B_2$
The Netherlands	Peanuts, peanut products	5(B_1)		Warenwet art. 3 quinquies Benelux Aanbeveling M(77)5-03.05.1977	Member of European Communities
	Fluid milk	0.1(M_1)		Administrative guideline of Public Health Inspectorate	
	Feeds	see EC		see EC	
New Zealand	All foods (import)	15			
Switzerland	Almonds, almond products, peanuts, peanut products, hazelnuts, hazelnut products, tree nuts, brazil nuts, pistachio nuts, coconut, sesame, poppy seed, pinetree pits, apricot and peach pits, cashew nuts, mixed nuts and tutti frutti	5 or 1 (B_1)	+	Kreisschreiben no. 21 des Eidgenossisches Gesundheitsamt 14.12.1977 and art. 6a Lebensmittel Verordnung	
	Milk and milk products	zero	+	art. 6a, Lebensmittel Verordnung	
USSR	All foods	5			
UK	Nuts and nut products	5(B_1)		Letter of Ministry of Agriculture Fisheries and Food mentioned in Special Circular 12/81, Leatherhead Food Research Association	Member of European Communities
USA	All foods and feeds	20		Section 402(a)(1) of the Food, Drug and Cosmetic act and supporting FDA Administrative Guidelines	

[a] Established tolerances given as the sum of aflatoxins B_1, B_2, G_1 and G_2 unless otherwise indicated.

[b] Some countries do not provide numerical tolerance or specify 'zero'. In both cases the maximum tolerated level is dependent on the detection limit of the analytical method.

Table 6.10. Maximum tolerated levels in various countries for mycotoxins other than aflatoxins (Schuller *et al.*, 1982)

Country	Commodity	Mycotoxin	Established tolerance (ng g^{-1})	Based on detection limit[b]	Remarks
Belgium	All foods	Patulin	zero	+	Koninklijk Besluit-03.01.1975
	All foods	Ochratoxin A	zero	+	Koninklijk Besluit-03.01.1975
	All foods	Sterigmatocystin	zero	+	Koninklijk Besluit-03.01.1975
	All foods	Zearalenone	zero	+	Koninklijk Besluit-03.01.1975
Denmark	Pork meats	Ochratoxin A	25		Limits established for kidney of animal slaughtered, calculated to achieve the levels in the other tissues
	Pork liver and kidney	Ochratoxin A	10		
Norway	Apple juices (conc.)	Patulin	50		Reconstituted basis
Sweden	Apple juices (conc.)	Patulin	50		Administered guideline
					Reconstituted basis
Switzerland	Juices (apple)	Patulin	50		Reconstituted basis
Canada	All grains for use in baby food	Deoxynivalenol	zero	+	Advisory to industry
	All other uses		indeterminate[a]		
USSR	All foods	Patulin Ochratoxin A Trichothecenes			

[a] Varies with circumstances of each crop and developing knowledge of deoxynivalenol toxicology.
[b] Some countries do not provide numerical tolerance or specify 'zero'. In both cases the maximum tolerated level is dependent on the detection limit of the analytical method used.

MOULD FERMENTATIONS

Many industrial processes use mould fungi for the production of enzymes, organic acids, antibiotics, flavour compounds and biomass or SCP. How safe are these products and what tests are carried out to ensure continuing safety? In all cases the producer organisms have been analysed for potential mycotoxin production and many of these fungi have been used for several decades (e.g. *Aspergillus niger* for citric acid production) without any observable toxic manifestations. Mould-derived enzymes such as amylases, proteinases destined to be used in the preparation of human foods or beverages must undergo stringent tests to ensure that toxic metabolites have not occurred during production and been carried through the purification processes. No important mycotoxins have been identified in preparations derived from acredited moulds.

When moulds are used for SCP processes in which the final product will be consumed then there must be long, exacting and costly toxicological and feeding tests before final official acceptance. *Fusarium* spp. have been used for the production of mycoprotein by Rank Hovis & McDougall and the final product, produced by a continuous fermentation process, has achieved the accolade of complete safety and acceptance and will soon become an acceptable food commodity.

Thus in all industrial mould fermentations, achievement of a mycotoxin-free product will only follow if certain criteria are strictly obeyed, viz.:

1. The use of mycotoxin-free substrates;
2. The use of pure cultures which have no known ability to produce mycotoxins;
3. The prevention of contamination by toxigenic moulds from the air and apparatus during processing, transportation and final storage of products.

MYCOTOXIN LEGISLATION

When food or feed is contaminated with a hazardous substance it becomes unfit for consumption. Legislation has been enacted in most countries imposing limits or tolerances on the concentration of such compounds and in most cases it includes mycotoxins. Countries vary in how they interpret and enforce such legislation. Consideration must be taken of geographic and agronomic characteristics, political strategy, state of industrialization and economic achievement. Since mycotoxins are occurring in foods and feeds mainly as natural contaminants and cannot in most cases be completely excluded without banning the susceptible products compromise decisions are usually made. Thus, countries with no domestic production of the susceptible commodities will normally have lower tolerances than those where the commodities are produced. Again where countries depend financially on an export market (e.g. peanuts and maize) the better quality commodity should meet the stricter legislation of the importing country but unfortunately leaving lower quality contaminated materials for local consumption and subsequent toxic effects.

The tolerated levels for mycotoxins have been arrived at by means of basic considerations. In the first instance the value must be the known non-toxic level of the mycotoxin and secondly must be at the minimum detectable level using the best analytical method available. The maximum tolerated levels of aflatoxins in foods and feeds is shown for several countries in Table 6.9 while Table 6.10 gives some limited information on other mycotoxins.

In the final answer there is no simple, assured way in which the safety level of a mycotoxin in a product can be determined. It is apropriate to end this Chapter by quoting Shank (1978): 'The most relevant question and the most difficult to answer is, what is to be done about mycotoxins in the human environment? Exposure must be minimized, of course, but to what level? At what level of mycotoxin should food be condemned or destroyed? The logical approach to these questions is to find the level at which the risk to human health is tolerable. Unfortunately, a scientific means of determining these levels is not available and who is to bear the burdensome responsibility of determining what risks to human health are tolerable? One must not quit in despair, but strive even more diligently, in the laboratory with animals and in the field with humans, to seek better answers to the problems of mycotoxins and human health'. (Note that since 1978 there have been considerable efforts made to replace animals with tissue cultures and human cell lines.)

FURTHER READING

Anon (1979). Mycotoxins, *Environmental Health Criteria*, **11**, World Health Organisation, Geneva.

Anon (1980). *Survey of Mycotoxins in the United Kingdom*, Fourth Report of the Steering Group in Food Surveillance, HMSO, London.

Bennett, G. A. and Shotwell, O. L. (1979). Zearalenone in cereal grains. *Journal of the American Chemical Society*, **56**, 812–819.

Brown, C. A. (1982). Aflatoxin M in milk. *Food Technology in Australia*, **34**, 228–231.

Bullerman, L. B. (1979). Significance of mycotoxins to food safety and human health. *Journal of Food Protection*, **42**, 65–86.

CAST (1979). Aflatoxin and other mycotoxins: an agricultural perspective. Council of Agriculture and Technology Report No. 80.

Davis, N. D. and Diener, U. L. (1978). Mycotoxins. In: *Food and Beverage Mycology*, L. R. Beuchat (Ed.), Avi Publishing, Connecticut, pp. 397–444

Hesseltine, C. W., Shotwell, O. L., Ellis J. J. and Stubblefield, R. D. (1966). Aflatoxin production by *Aspergillus flavus*. *Bacteriological Reviews*, **30**, 795–805.

Jarvis, B. (1976). Mycotoxins in food. In: *Microbiology in Agriculture, Fisheries and Food*, F. A. Skinner and J. G. Carr (Eds), Academic Press, London, pp. 251–267.

Jarvis, B. (1982). The occurrence of mycotoxins in UK foods. *Food Technology in Australia*, **34**, 508–514.

Kiermeier, F. (1979). Einsichrung in die Mykatoxin—Problematik. *Zeitschrift fur Lebensmittel Untersuchung Forschung*, **167**, 118–127.

Krogh, P. and Neisheim, S. (1982). Ochratoxin A. In: *Environmental Carcinogens: Selected Methods of Analysis*, vol. 5, Some Mycotoxins, H. Egan, L. Stoloff, M. Castegnaro, P. Scott, I. K. O'Neill and H. Bartsch (Eds), International Agency for Research on Cancer, France, pp. 247–253.

Northolt, M. D., Van Egmond, H. P., Soentoro, P. and Deijll, E. (1980). Fungal growth and the presence of sterigmatocystin in hard cheese. *Journal of the Association of Official Analytical Chemists*, **63**, 115–119.

Norton, D. M., Toule, G. M., Cooper, S. J., Partington, S. R. and Chapman, W. B. (1982). The surveillance of mycotoxins in human food. In: *Proceedings 4th Meeting Mycotoxins in Animal Disease*, G.A. Pepin, D.S. Patterson and D.E. Gray (Eds), Ministry of Agriculture, Fisheries and Food, Lion House. Northumberland.

Osborne, B. G. (1982). Mycotoxins and the cereal industry—a review. *Journal of Food Technology*, **17**, 1–9.

Rodricks, J. V., Hesseltine, C. W. and Mehlman, M. A. (Eds). (1977). *Mycotoxins in Human and Animal Health*, Pathotox, Illinois.

Schuller, P. L., Stoloff, L. and Van Egmond, H. P. (1982). Limits and regulations. In: *Environmental Carcinogens — Selected Methods of Analysis*, vol. 5, *Some Mycotoxins*, H. Egan, L. Stoloff, M. Castegnaro, P. Scott, I. K. O'Neill and H. Bartsch (Eds), International Agency for Research on Cancer, France, pp. 107–116.

Shank, R. C. (1978). Mycotoxicosis of man: dietary and epidemiological considerations. In: *Mycotoxic Fungi, Mycotoxins, Mycotoxicoses*, T. D. Wyllie and L. G. Morehouse (Eds), vol. 3, Marcel Dekker, New York, pp. 1–9.

Shotwell, O. L. (1977). Assay methods for zearalenone and its natural occurrence. In: *Mycotoxins in Human and Animal Health*, J. V. Rodericks, C. W. Hesseltine and M. A. Mehlman (Eds), Pathotox, Illinois, pp. 403–413.

Shotwell, O. L. and Hesseltine, C. W. (1983). Five-year study of mycotoxins in Virginia wheat and dent corn. *Journal of the Association of Official Analytical Chemists*, **66**, 1466–1469.

Stoloff, L. (1976). Occurrence of mycotoxins in foods and feeds. In: *Mycotoxins and other Fungal Related Food Problems*, J. V. Rodricks (Ed.), American Chemical Society, Washington, pp. 23-50.

Stoloff, L. (1980a). Aflatoxin M in perspective. *Journal of Food Protection*, **43**, 226–230.

Stoloff, L. (1980b). Aflatoxin control: past and present. *Journal of the Association of Official Analytical Chemists*, **63**, 1067–1073.

Thiel, P. G., Marasas, W. F. O. and Meyer, C. J. (1982). Natural occurrence of *Fusarium* toxins in maize from Transkei. In: *Mycotoxins and Phycotoxins*, Vth IUPAC Symposium, Vienna, pp. 126–129.

Ueno, Y. (1983). *Trichothecenes: Chemical, Biological and Toxicological* Aspects, Elsevier, Amsterdam (several chapters on natural occurrence).

Vesonder, R. F. and Hesseltine, C. W. (1981). Vomitoxin: natural occurrence on cereal grains and significance as a refusal and emetic factor to swine. *Process Biochemistry*, (1980/1981), Dec/Jan, 12–16.

Watson, D. H. (1984). Survey and control of mycotoxins in animal and human foods. *Chemistry and Industry*, 536–540.

Yokotsuka, T. and Sasaki, M. (1982). Risks of mycotoxins in fermented foods. *6th International Symposium on Fermentation Technology*, Pergamon Press, London, Ontario.

CHAPTER 7

Mycotoxin Analyses

Many mycotoxins are now recognized to be involved in the aetiology of certain human and animal diseases. An awareness of the levels of contamination of mycotoxins in natural products can only be obtained by developing good analytical methodologies for detecting mycotoxins in foods, mixed feeds and feed ingredients, animal tissues, blood, urine and milk.

Since mycotoxins display a wide diversity of chemical structure, there are no uniform methods of analysis either collectively or for a specific toxin in various foods or feeds. However, the main mycotoxins can now be readily identified qualitatively and quantitatively and most current investigations concentrate on increasing sensitivity, accuracy and reproducibility and above all to decrease the time of analysis. At present satisfactory limits of detection can be achieved for most mycotoxins of interest and these limits are, generally, adequate for legislative, veterinary and medical requirements.

The most practical methods for detection are physico-chemical in nature while biological assays have varying levels of utilization and importance. Methods are available for rapid screening of samples for many toxins. Such multi-mycotoxin methods can be introduced at an intermediate stage in the screening of potentially contaminated samples and only those samples showing positive presence of toxin(s) need then be subjected to the methods specifically designed for the quantitative analysis of a particular mycotoxin.

The analytical procedures for detecting mycotoxins will generally follow the same flow pattern of: sampling, extraction, clean-up, separation, detection and quantitation and finally confirmation (Table 7.1).

SAMPLING

Mycotoxins are rarely uniformly distributed throughout natural products, rather their occurrence will be uneven and spasmodic (Chapter 6). Mycotoxins are generally found in high concentrations at the sites where the toxigenic fungi have invaded the product. When analysing large quantities or lots of agricultural products, e.g. peanuts, cereal grains, compounded animal feeds, care must be exercised to achieve a situation where the final samples for analysis can be as truly representative of the whole lot as possible. Excellent guidelines have been drawn up to satisfactorily achieve accurate sampling in growing crops, at harvesting, transport, processing and ultimate utilization (Davies *et al.*, 1980).

Table 7.1. Basic steps in chemical analyses of mycotoxins (after Romer, 1976)

Step	Description	Purpose
1. Sampling	Probe of automatic sampler	Representative sample
2. Sample preparation	Grinding, mixing, subsampling	Representative sample
3. Extraction	Shaker or blender	Separate the toxin from compounds insoluble in the extraction solution
4. Clean-up	Liquid–liquid partitioning (separatory funnel)	Separate the toxin from groups of compounds in the sample extract
	Column chromatography	
	Divalent metal clean-up (Pb^{2+}, Fe^{2+}, Cu^{2+})	
5. Final separation	Thin layer chromatography (TLC)	Separate the toxin from remaining compounds in the sample extract that might interfere with the toxin
	Gas–liquid chromatography (GLC)	
	Liquid–liquid chromatography (LC)	
	Minicolumn chromatography	
6. Detection and quantitation	Fluorescence on TLC plate	Detection and measurement of response
	Fluorescence in solution	
	UV absorption in solution	
	GLC-flame detector	
7. Confirmation	TLC separation and detection of derivative of mycotoxin	Identification of chemical compound
	Biological test	
	Mass spectrometry	

The main difficulty in sampling for mycotoxins arises from the heterogeneity of toxin distribution in contaminated unprocessed commodities. The larger the individual particle or seed the greater will be the problem of selection. Thus, for example, a batch of peanuts of 10 tonnes containing about 2×10^7 kernels may contain only a few highly contaminated nuts. In terms of sampling probabilities what will be the chance (P_0) of a randomly selected sample not containing contaminated kernels? When the contamination ratio for a batch of peanuts is 10^{-6} (i.e. a 10 tonne batch will contain about 20 affected kernels) then a 4 kg sample comprising 8×10^3 kernels will have an expected number of contaminants $\lambda = 8 \times 10^{-3}$ and $P_o = \exp(-\lambda) = 0.992$. This means that about 99 per cent of the time 4 kg withdrawn samples will not contain contaminated kernels (Brown, 1982).

Even then it can be seen that when the contamination rate is low larger samples must be used. In practice, large sample assessments are achieved by grinding up large samples and subsampling the comminuted material. The total variance of the analytical procedure is mainly caused by sampling variability, while subsampling variability and analytical variability are more or less independent of toxin concentrations (Fig. 7.1) (Whitaker, 1977). Sampling variability arises from the variability of toxin levels in the kernels selected in the sample. The ideal sample should be a continuous sample taken during production or by thoroughly mixing probe samples drawn from widely dispersed bags (Fig. 7.2). Examples of sampling procedures used by the United States Food and Drug Administration are shown in Table 7.2. The expected error in estimating the average mycotoxin concentration in a population of lots is directly proportional to the variance in mycotoxin concentration among the population of lots and inversely proportional to the number of lots sampled.

Absolute recommendations on sample size are not feasible but must be related to the size of the specific particles and level of homogeneity. In this way larger samples must be taken with increasing particle size, e.g. peanuts > corn > wheat > rice > milled products. Although larger samples will undoubtedly increase accuracy there will be commensurate cost increases. Whitaker and Dickens (1979) calculated the effect of sample size on sample distribution and total batch distribution for mycotoxin distribution. Using shelled corn as an example, 5 kg samples were shown to be adequate for most survey purposes.

In practice the original coarse samples should be ground to pass through a No. 14 sieve, thoroughly blended and properly subdivided to a 1 kg sample. The entire 1 kg sample should then be ground to pass through a No. 20 sieve, thoroughly blended and properly subdivided to 50 g analytical samples. The sampling of small grains, oilseed cake, foods and feeds is also problematic although the distribution of the mycotoxins is generally not as uneven as with groundnuts and corn. Fluids and well-mixed process products such as milk, milk products, beer, cider, etc., do not normally present such sampling problems.

As far as possible, samples should be ground and subsampled immediately after collection. When subsamples cannot immediately be analysed, they should be stored under refrigeration or dried.

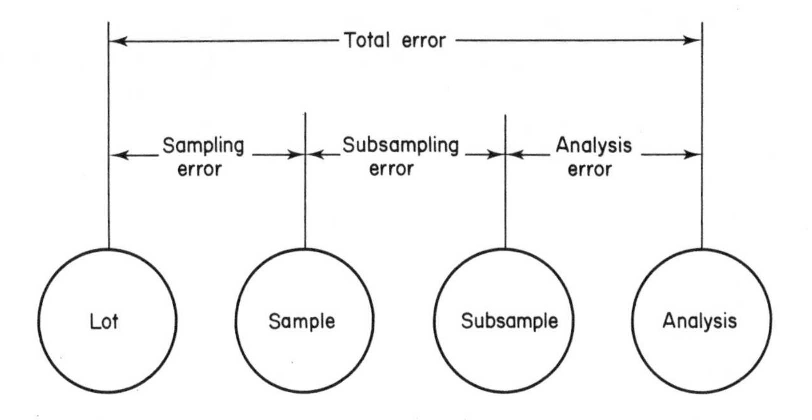

Fig. 7.1. Types of error associated with mycotoxin tests

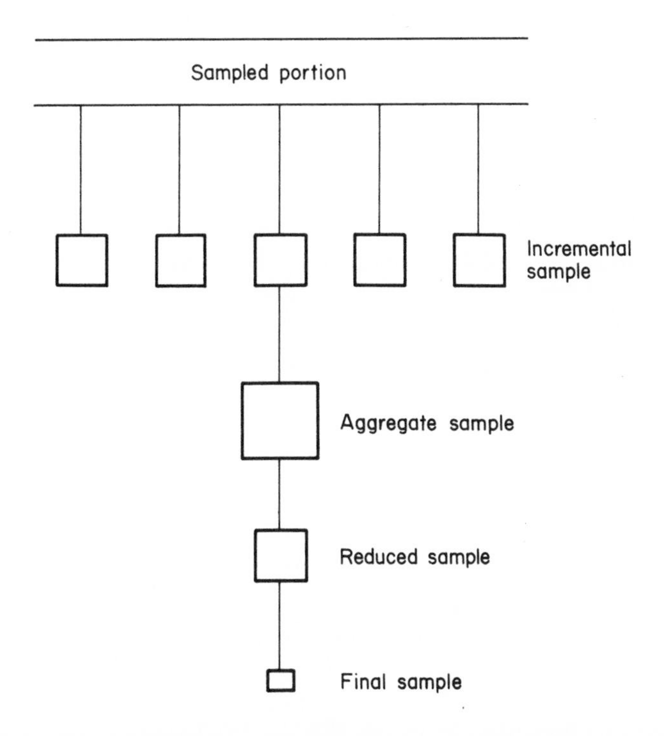

Fig. 7.2. Sampling method for foodstuffs according to EC Directive

Table 7.2. Sampling procedures for mycotoxin analyses[a]

Product	Package type[b]	Lot size	Number of sample units[c] (minimum)	Unit size (minimum)	Total sample size (minimum)
Peanut butter : smooth	Consumer or bulk	NA	24 12	8 oz 1 lb	12 lbs 12 lbs
Peanut butter : crunchy		NA	48	1 lb	48 lbs
Peanuts : shelled, roasted or unroasted					
Peanuts : ground for topping					
Brazil nuts : in shell (in import status)	Bulk	<200 bags 201-800 bags 801-2000 bags	20 40 60	1 lb 1 lb 1 lb	20 lbs 40 lbs 60 lbs
Corn : shelled, meal flour or grits	Consumer or bulk	NA	10	1 lb	10 lbs
Cottonseed	Bulk	NA	15	4 lbs	60 lbs
Oilseed meals : peanut meal, cottonseed meal	Bulk	NA	20	1 lb	20 lbs
Milk : whole, skim, low fat	Consumer Bulk	NA NA	10 –	1 lb –	10 lbs 10 lbs
Small grains : wheat, sorghum, barley, etc.	Bulk	NA	10	1 lb	10 lbs
Dried fruit[d] : e.g. figs	Consumer or bulk	NA	50	1 lb	10 lbs
Mixtures containing commodities susceptible to mycotoxin contamination	Consumer	NA			
Commodity particles relatively large			50	1 lb	50 lbs
Commodity particles relatively small			10	1 lb	10 lbs

[a] Reproduced from Inspection Operation Manual of the Food and Drug Administration, United States Department of Health, Education and Welfare (dated 10/19/79).
[b] Containers for samples of unprocessed, intact nuts, seeds or grains must be sufficiently porous to provide for dissipation of moisture produced by respiration of the nut, seed or grain.
[c] To be collected from as many random sites in the lot as possible.
[d] Optional sampling programme for seeds or dried fruit with a low incidence of contamination: take initial 10 × 1 lb sample. If any aflatoxin is detected, resample 50 × 1 lb sample for a determination of contamination level on which to base a regulatory judgement.
NA : not applicable.

EXTRACTION AND CLEAN-UP

Mycotoxins normally occur in extremely low concentrations (ppb–ppm) in the complex chemical milieu of organic material. Before these mycotoxins can be measured, they must first be extracted from the contaminated substance and co-extracted chemicals substantially removed by clean-up procedures. Initial transfer of the mycotoxins and co-extractives into an extracting solvent(s) involves homogenization either for a short period (1–3 min) in a high shear-rate blender or for longer periods (around 30 min) by hand or mechanical shaking in a flask.

In general, mycotoxins are soluble in slightly polar solvents and usually insoluble in completely non-polar solvents. Mycotoxins may exhibit differential binding to organic molecules and differing degrees of solubility in water. In practice, mycotoxins are extracted using mixtures of organic solvents such as chloroform, acetonitrile, methanol, acetone, ethylacetate or dichloromethane, often in combinations with small amounts of water or acids. Aqueous solvents more easily penetrate hydrophilic tissues and enhance toxin extraction. With the correct proportions of water to solvent the toxins are often more readily partitioned into the solvent.

The presence of pigments, fats and lipids in extracts will seriously reduce the efficiency of ensuing separation techniques. By adding fat solvents such as hexane to the extraction solvents many of the fats and lipids can be partitioned into the hexane portion of the solvent and discarded. When the solvents used in the extraction are immissible, partitioning is best done in a glass separatory funnel and the separating layers removed in sequence.

Although many interfering compounds may be partially removed during the extraction sequence further clean-up of the extract is normally necessary. Column chromatographic techniques for clean-up are now widely practised and the choice of column packing depends on the substances to be eluted and the particular mycotoxins. The types of adsorbents used include silica gel, polyamide, Florisil, Sephadex or aluminium oxides. The sample extracts are applied to the clean-up columns and, after elution of the column with suitable solvents that do not elute the mycotoxin, appropriate elution solvents can be applied to the column and the mycotoxin eluted and collected. Many types of solvents and solvent mixtures have been used to selectively remove non-lipid interfering substances from adsorbent columns. Columns may either be prepared in the laboratory or purchased as prepacked cartridges. Microprocessor controlled clean-up systems are routinely employed where large numbers of samples are being used yielding highly purified extracts for further analysis.

The ultimate aim of the clean-up procedure is to remove the greater part of the co-extracted substances thus reducing the chemical complexity of the final extract which can then proceed to the detection and quantitation phases of the analytical method.

The final stage in the preparation of the samples for analysis is volume reduction either by evaporating the solvent in a rotary evaporator under reduced pressure or on a steam bath or enclosed hot plates (under a stream of nitrogen). The dried sample can then be redissolved in a known volume of solvent to be used for the analysis. Some of the most widely used solvent combinations for extracting the main naturally occurring mycotoxins are given in Table 7.3.

Table 7.3. Summary of extraction solvents used in Official AOAC Methods of Analysis for several mycotoxins

Toxin	Commodity	Extraction solvent
Aflatoxins	Corn, cottonseed	Acetone : water (85 : 15)
	Green coffee beans, soybeans, coconut, copra, copra meal	Chloroform : water (91 : 9)
	Cocoa beans	Defat with hexane then chloroform
	Peanut products, pistachio nuts	Chloroform : water (91 : 9) or methanol : water (55 : 45) plus hexane (39 : 32 + 29)
	Powdered milk	Acetone : water (70 : 30)
Ochratoxins	Barley	Chloroform + 0.1 M phosphoric acid
Patulin	Apple juice	Ethyl acetate
Sterigmatocystin	Barley, wheat	Acetonitrile : 4% potassium chloride (9 : 1)
Trichothecenes	Cereals	Methanol : water (9 : 1)

ANALYTICAL METHODS

In the development of analytical methods for mycotoxins the use of chemical detection has been preferred to biological assay. Chemical assays are more easily quantifiable, less subject to interference by non-fungal food co-extractives and normally more sensitive than biological assays.

Chromatographic separation

Although final extracted samples have been subjected to clean-up procedures they will still normally contain large amounts of co-extracted substances that can interfere with most chromatographic techniques. The chromatographic methods most widely and routinely used are one- and two-dimensional thin-layer chromatography (TLC) and high performance liquid chromatography (HPLC). Since most mycotoxins are non-volatile, gas–liquid chromatography (GLC) has limited use but has become particularly important with the non-fluorescing trichothecene mycotoxins.

Thin-layer chromatography

Thin-layer chromatography has been the most widely used analytical method for separating and identifying mycotoxins from concentrated extracts. The technique of TLC involves coating a glass plate with silica gel, applying a concentrated sample on a base line, separation by solvent migration, drying and characterization of the resultant spots.

With every combination of solvents each mycotoxin will have a characteristic migration and separation pattern known as the R_f value. A vast combination of solvent mixtures has been investigated and a worthwhile introduction to broad solvent categories for mycotoxin separation can be found in the Official Methods of the Association of Official Analytical Chemists (AOAC) (Table 7.3). TLC methods are less often used definitively for the trichothecene mycotoxins due to their lack of a reactive group or an applicable group for detection of the molecule. However, TLC is often used for the trichothecenes as a preparative stage of separation or clean-up prior to using gas–liquid chromatography. Migration data for mycotoxins on thin-layer chromatograms are only approximate and all unknown samples must be compared to reference standards on the same plate.

Recent innovations in TLC analytical techniques for mycotoxins include two-dimensional chromatography in which the sample is developed in one direction with a given solvent, dried and then developed in a second direction, perpendicular to the first, with a second solvent. Two-dimensional chromatography is particularly suitable for sample extracts containing large amounts of co-extracted substances. Thus, development in the first direction serves as a clean-up step while the second direction is for the actual detection/quantitation.

Multimycotoxin analysis using TLC

When assaying biological materials for mycotoxins it is seldom possible to know in advance which mycotoxins, if any, will be present. For this reason many multi-mycotoxin screening techniques have been developed which permit simultaneous analysis of several mycotoxins. In this way only extracts which show a positive presence need then be subjected to methods specifically designed for the quantitative analysis of the specific mycotoxin. TLC techniques have formed the basis of most multi-mycotoxin studies. These methods are of particular importance when screening cereals and cereal foodstuffs, oil seeds, straight and compounded animal feedstuffs, animal laboratory diets and dairy concentrates.

One example of a well-proven multi-mycotoxin screen will be outlined to indicate the type of flow pattern used when dealing with mould-contaminated biological material (Gorst-Allman and Steyn, 1979). The suspect sample is extracted in a Waring blender with methanol-chloroform (1:1) (400 ml) at 23 000 rev min^{-1} for 4 × 1 min. The mixture is filtered and the filtrate evaporated to dryness. The resultant brown residue is partitioned between n-hexane and 90 per cent methanol (1:1) (200 ml), the *n* -hexane layer discarded and the methanol layer evaporated to dryness. The brown solid is partitioned between chloroform and water (1:1) (200 ml) and the chloroform layer extracted with saturated sodium hydrogen carbonate solution (3 × 100 ml). The chloroform layer is concentrated and contains the so-called neutral mycotoxins (viz. aflatoxin B_1, sterigmatocystin, zearalenone, patulin, T-2 toxin, roquefortin, penitrem A, fumitremorgen B and roridin A) that were present initially. The aqueous layer is then carefully acidified to pH 2 (with 0.5 N HC1) and extracted with chloroform (3 × 100 ml). The chloroform extract is concentrated and contains any of the so-called *acidic* mycotoxins (ochratoxin A, citrinin, α-cyclopiazonic acid and penicillic acid) that may be present. Separation of the various mycotoxins in each category can be achieved with many different solvent combinations and the R_f values calculated on the basis of known mycotoxin standards co-chromatographed.

Obviously this method does not give good clear separation of all the mycotoxins due to some degree of overlapping or tailing off of some of the mycotoxins but it does allow preliminary evidence of presence or absence of specific mycotoxins. If any toxin is detected then further analysis can be done using a more accurate and appropriate extraction and clean-up method for that toxin. Most multi-mycotoxin detection procedures must be considered primarily as screening techniques since they can rarely be considered quantitative and are not normally suitable for confirmation of identity, but are rapid in terms of total information gained.

Minicolumn detection

In agricultural practice there is a need for an economical and practical screening method to analyse large numbers of food and feed samples most of which will be mycotoxin-free. In particular the method should be suitable for use by relatively inexperienced field or factory workers. Minicolumn detection methods have come to occupy an important niche in rapid screening methodology and have been used for the detection of total aflatoxins in almonds (about 5 $\mu g\ kg^{-1}$), corn, peanuts, cottonseed (about 10 $\mu g\ kg^{-1}$), and mixed feeds (about 15 $\mu g\ kg^{-1}$).

The Velasco minicolumn consists of a glass tube (15 cm × 6 mm) with a glass wool plug at the base covered sequentially with a layer of calcium sulphate, a layer of Florisil, a layer of silica gel, a layer of alumina and finally a layer of calcium sulphate (Fig 7.3). The test sample is added at the top of the column and eluted with chloroform:acetone (9:1). The coloured impurities present in most samples will be removed by the alumina while

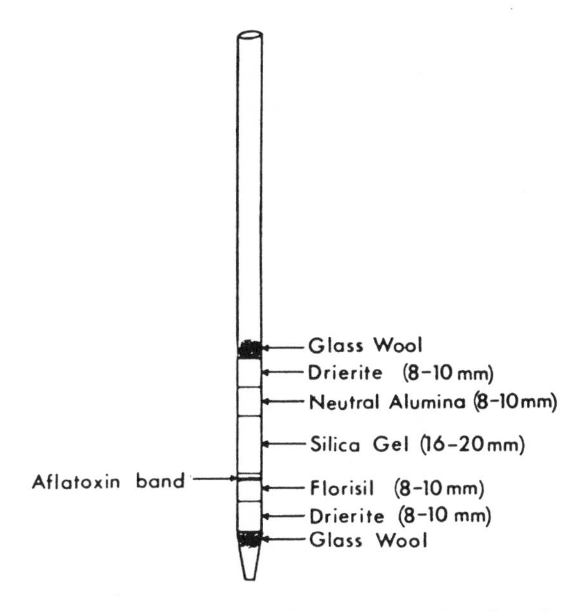

Fig. 7.3. Diagram of minicolumn used for aflatoxin detection

the calcium sulphate acts as a drying agent. The aflatoxin moves through the column becoming firmly attached to the Florisil. Detection is again under ultra violet (UV) lamp or in a chromatic cabinet. Levels as low as 5 ppb can be detected and a positive result should then be verified by conventional quantitative methods for aflatoxin. Minicolumn methods are now widely available and used for detection of aflatoxin, ochratoxin A and zearalenone.

Visualization and quantitation

After satisfactory separation of the extracts has been achieved by TLC methods the mycotoxins must be detected and measured. Mycotoxins are detected by the physico-chemical properties of the molecule often involving a combination of absorption and fluorescence properties or the use of chromogenic reagents. Many mycotoxins such as the aflatoxins absorb UV light. Not only do the aflatoxins absorb UV light but they also re-emit part of the energy of the absorbed UV light as visible light (i.e. they fluoresce) and in this manner the mycotoxins can be easily detected. The intensity of fluorescence is a measure of the concentration of the toxin and can be determined visually or more accurately with a densitometer and compared to known standards. The B and M aflatoxins fluoresce blue while the G toxins fluoresce green and ochratoxin A greenish-blue; sterigmatocystin fluoresces dull brick red when exposed to long-wave UV light while zearalenone fluoresces a bluish-green in short-wave UV light. Patulin and penicillic acid, while not fluorescing in long-wave UV light, can be made to fluoresce by exposure to ammonia fumes or by spraying with 3 per cent aqueous ammonium hydroxide, patulin fluorescing pale blue and penicillic acid bright intense blue (Table 7.4). Visualization procedures for the trichothecenes have been developed by making use of reactions at the 12,13-epoxy group in the trichothecene nucleus. In this method 4-(*p* -nitrobenzyl) pyridine is sprayed onto the plate, heated for 30 minutes at 250° C and then sprayed with tetraethyl pentamine. The trichothecenes show up as blue spots on a light blue background (Table 7.5). Reference standards for most mycotoxins are available from many commercial and government sources (Egan *et al.*, 1982 for details).

Table 7.4. Visualization techniques used for detection of mycotoxins in feedstuff extracts (Steyn, 1980)

Mycotoxin	Visualization of spots	Interpretation at	Colour of spot
Aflatoxin B_1		360 nm	Blue fluor.
Ochratoxin A	Treatment with NH_3 vapour (10 min)	360 nm	Blue fluor.
Patulin	Spraying with MBTH solution[a] followed by heating for 15 min at 110° C	360 nm	Yellow fluor.
Sterigmatocystin	Spraying with $A1C1_3$ solution followed by heating for 10 min at 110° C in oven (20 g $A1C1_3$ in 100 ml ethanol)	360 nm	Yellow-green fluor.
Zearalenone	Before spraying with $A1C1_3$ solution for sterigmatocystin	254 nm	Blue-green fluor.
Penicillic acid	Treatment with NH_3 vapour (10 min), followed by heating plate for 5 min at 110° C in oven	360 nm	Blue fluor.
Citrinin		360 nm	Yellow fluor.
Cyclopiazonic acid	Spraying with Ehrlich reagent[b]	Daylight	Violet

[a] MBTH-solution: 0.5 g 3-methyl-2-benzothiazoline hydrazone hydrochloride in 100 ml H_2O.
[b] Ehrlich reagent: 2 g *p*-dimethylaminobenzaldehyde in 100 ml 10% HC1.

The presence of a spot on a TLC plate can only be considered as presumptive evidence of identity and further confirmatory tests are required. Confirmation may include re-chromatographing the extract with a number of different systems, e.g. varying the solvent systems, support or switching between TLC and HPLC, while for the aflatoxins, confirmation can be achieved by spraying the plate with 25 per cent sulphuric acid which will change the fluorescence colour of the aflatoxin spots to yellow. However, some other compounds may also become yellow so this method although useful is not absolute. Formation of chemical derivatives and subsequent comparison of mobilities and fluorescent properties on TLC with standard derivatives can be quite conclusive. The ultimate and best proof of identity will be by mass spectroscopy.

Table 7.5. TLC visualization procedures for trichothecenes (Gilbert, 1984)

Procedure	Trichothecene	Limit of detection[a] (μg per spot)	Colour
p-Anisaldehyde (MeOH, acetic acid, H_2SO_4 soln.)	Deoxynivalenol	0.05	Yellow
	Diacetoxyscirpenol	0.10	Purple
	T-2 toxin	0.10	Brown
	HT-2 toxin	0.20	Brown
20% H_2SO_4 soln.	Deoxynivalenol	0.05	Yellow
	Diacetoxyscirpenol	0.20	Purple
	T-2 toxin	0.20	Grey
	HT-2 toxin	0.50	Grey
10% Aluminium chloride	Deoxynivalenol	0.10	Blue (fluor)
	Nivalenol	0.10	
	Fusarenon-X	0.10	
4-(*p*-Nitrobenzyl) pyridine	All trichothecenes	0.02–0.2	Blue spots
Nicotinamide/2-acetylpyridine	All trichothecenes	0.02–00.05	Light blue (fluor)

[a] Determined as pure reference standards.

High performance liquid chromatography

High performance liquid chromatography (HPLC) is a separation technique that has become increasingly used for the analysis of mycotoxins because it offers increased sensitivity and improved accuracy over TLC methods.

The technique of HPLC involves the separation of the various constituents of a sample followed by their individual detection and measurement. Separation is achieved by a competitive distribution of the sample between two phases, one a mobile liquid and the other a stationary liquid or solid. The stationary phase is supported in a column about 25 cm long by 4 mm internal diameter. High separating efficiency has been achieved by the optimization of column parameters and, in particular, the particle size in the column packing. The mobile phase is forced by the pump through the separatory column and then flows through the detector. As each sample component passes through the detector a change in electrical output is produced which is recorded on a moving chart to give a chromatogram. The time taken for a substance (mycotoxin) to pass through the column under fixed conditions is constant and is called the retention time. Comparison of retention times with those of standards enables HPLC to be used for qualitative analysis. The area under each peak on the chromatogram is proportional to the component concentration and this enables HPLC to be used for quantitative analysis.

HPLC coupled to sensitive detection and sophisticated data retrieval has facilitated the identification of selected mycotoxins and their quantification at levels much lower than with TLC. Sensitive methods have now been developed for many mycotoxins. Aflatoxins because of their fluorescence can be coupled to fluorescence detectors and limits of detection as low as 1 $\mu g\ kg^{-1}$ can be achieved. Sterigmatocystin which is UV absorbent can be analysed by HPLC/UV detector and levels as low as 1 $\mu g\ kg^{-1}$ are possible but sample interference can be a problem.

For many mycotoxins HPLC techniques have proved a good alternative to TLC especially where extremely low levels of the mycotoxins need to be detected, e.g. aflatoxin M_1 in milk and milk products.

Gas–Liquid chromatography

Gas–liquid chromatography (GLC) is a separation technique applicable to compounds that exert significant vapour pressure at temperatures below those of excessive pyrolysis. Such compounds can be converted to stable, volatile derivatives that can be separated by vapour-phase chromatography. Since most mycotoxins are non-volatile this method of analysis has not been widely used. However, for the trichothecene mycotoxins which do not lend themselves easily to TLC and HPLC techniques, GLC combined with mass spectrometry has been a most effective method for identification and quantitation.

In GLC a glass or metal column about 200 cm × 2 mm internal diameter is packed with a suitable absorbent such as 80–100 mesh Chromosorb W. The sample is injected and flash-evaporated at one end of the column, the entire length of which is maintained at an elevated temperature (170 to 225° C). The volatilized derivatives are swept through the column by a stream of an inert gas, e.g. argon, helium or nitrogen flowing at a constant rate. Each fraction of the mixture moves on the column at a different rate determined by its ratio of partition between the gas phase and the non-volatile liquid (stationary) phase. The presence of the mycotoxin derivatives in the gas emerging from the column can be detected by physical or chemical means and the data automatically recorded on a chart as a series of peaks. As with HPLC the area under each peak is proportional to the concentration of the particular component in the mixture. Closely similar derivatives can be identified from the time necessary to elute a component from the column (retention time).

The best technique for the quantitation of trichothecene mycotoxins is GLC with electron capture (EC) or mass spectrometric (MS) detection. Sample preparation will normally involve extensive clean-up by column chromatography on a silica gel. Trichothecenes as such are not sufficiently volatile for direct analysis by GLC. Rather they must be derivatized through free hydroxyl groups on the molecules to form trimethyl-silyl (TMS) ethers or heptafluorobutyryl (HFB) esters that will be sufficiently volatile for analysis by GLC.

Although detection of trichothecenes by GLC flame ionization is much more sensitive than TLC, with a limit of about 1 μg per component on a column, assessment of the identity of peaks can be problematic because of the complexity of biological material. For unequivocal confirmation of trichothecenes mass spectrometry is required and will require approximately 10–20 μg of each trichothecene injected on columns equating to a contamination level of about 100–200 $\mu g\ kg^{-1}$ after a normal clean-up procedure.

GLC with electron capture (GLC-EC) detection and confirmation with mass spectrometry has been successfully used for identifying trichothecenes, particularly vomitoxin, in United States and Canadian surveys of wheat. This method has a low detection limit (10 ppb), is reliable and has been tested on a large number of contaminated samples.

Biological Methods

Biological Assays

The ultimate test of toxicity of a mycotoxin is to experimentally test the effect of the suspected toxin on a living system. Since mycotoxin effect will be related to body weight of the test organism considerable efforts have been made to find meaningful biological systems that will give accurate, cheap and rapid evidence of toxic properties.

One-day old ducklings. In the earliest biological assays one-day old ducklings were used to test for the presence of aflatoxins in suspect food and feed samples. The toxic samples were introduced into the gizzard of the ducklings by way of a plastic tube. With high dosages the ducklings died during the following seven days. The acute LD_{50} values for the aflatoxins have been calculated in $mg\ kg^{-1}$ as:

Aflatoxin B_1, 0.36; B_2, 1.69; G_1, 0.78 and G_2, 2.45

Chronic and sublethal levels of toxicity can cause proliferation of the bile duct and can be used as a semi-quantitative index. Detectable biliary proliferation can be observed with aflatoxin B_1 at concentrations as low as

0.4 μg day^{-1} for five days. This test has also been successfully used to detect aflatoxin M_1 in both liquid and powdered milk.

Chick embryo. The only accepted bioassay method in the *AOAC Manual of Methods* is the chick embryo bioassay for aflatoxin B_1 but it has also been used for many other mycotoxins. It is recommended that eggs from fertile, inbred single-comb White Leghorns are used although other types may also be considered. The eggs can be injected prior to incubation or several days after incubation when the presence of a live embryo can be assured. The test extract can be injected into either the air sac or into the yolk sac of the fertile egg and dose rates as low as 0.3 mg of aflatoxin B_1 can be lethal.

Brine shrimp larvae. The brine shrimp *Artemia salina* has been widely used as a convenient and cheap means of assaying for mycotoxins. The eggs of this sea creature are easily stored in laboratory conditions and when placed in saline water at 27° C will hatch out in 24 h. The larvae are phototrophic and by this means can be easily separated from the eggs. Between thirty and fifty larvae in a small volume of water are tranferred and allowed to incubate at 37.5° C for 24 h and the percentage dead calculated. The larvae have been shown to be sensitive to aflatoxin B_1, ochratoxin A, gliotoxin, sterigmatocystin and several trichothecenes.

Fertilized eggs of the sea urchin *Hemicentratus pulcharimus* have been used to screen for several mycotoxins, in particular, the trichothecenes.

Skin toxicity tests. Skin toxicity tests have been widely used for screening toxigenic fusaria and related genera. The test extracts are applied to closely cropped skin of certain animals such as rabbit, guinea pig or rat and the skin reaction noted. The main features are erythema, oedema and necrosis. From 5 to 60 μg ml^{-1} of T-2 toxin and diacetoxyscirpenol can be quantitatively estimated by cutaneous injections of rats and rabbits.

Vomiting. Many farm animals will reject or refuse feed contaminated with *Fusarium* fungi. A mouse bioassay method has been developed and using this method, feed refusal and vomiting have been shown to be clearly associated with feed contaminated with vomitoxin, diacetoxyscirpenol or T-2 toxin. It is believed that these trichothecenes stimulate the chemoreceptor trigger zone in the medulla oblongata. One day-old ducklings can also be used for this bioassay.

Cytotoxicity tests. Certain mycotoxins, especially the trichothecenes, possess a potent cytotoxicity to eukaryotes and several cell culture systems have been developed for rapid and specific bioassays (Table 7.6).

Microorganisms. Attempts have been made to utilize potential growth inhibition by mycotoxins of microorganisms, particularly bacteria, as a rapid means of measuring toxicity. Known quantities of the mycotoxin are added to paper discs and placed on nutrient agar seeded with a suspension of the test organism, e.g. *Bacillus brevis*, *B. stearothermophilus* or *Escherichia coli*. The *Bacillus* spp. appear to have the greatest potential for sensitive bioassay of mycotoxins. With such systems a standard curve may be constructed similar to those used for antibiotic assays and from which quantitative results can be obtained. However, these methods have not yet found wide practice.

Table 7.6. Cytotoxicity of trichothecenes to cultured cells (Ueno, 1983)

		LD_{50} ($\mu g\ ml^{-1}$)		
Types	Trichothecenes	HeLa	HEK	HL
A	Trichodermol	5.0	3.0	2.0
	Monoacetoxyscirpenol	0.1	0.1	0.3
	Diacetoxyscirpenol (DAS)	0.01	0.01	0.001>
	Neosolaniol	0.1	0.06	0.05
	Acetylneosolaniol	0.3	1.0	0.1
	7,8-dihydroxy-DAS	0.3	0.2	0.3
	T-2 toxin	0.01	0.01	0.003>
	HT-2 toxin	1.0	0.1	0.01
	Acetyl T-2 toxin	1.0	0.8	0.03
	Calonectrin	3.0	0.8	0.03
	Deactylcalonectrin	7.0	0.0	1.0
B	Nivalenol	0.3	1.0	0.3
	Fusarenon-X	0.1	1.0	0.3
	Deoxynivalenol	1.0	3.0	0.5
	Monoacetyldeoxynivalenol	10.0	10.0	10.0<
	Trichothecin	0.1	0.1	0.1
	Tetraacetylnivalenol	10.9<	10.0<	10.0
C	Crotocin	0.5	0.6	2.0
D	Verrucarin A	0.005	0.002	0.0003>
	Roridin A	0.0003	0.0003	0.0003>

Trout. Among the larger animals the trout is one of the most sensitive to aflatoxin and exhibits characteristic pathological changes of the liver. However, maintenance of the trout can be difficult and expensive.

Several other biological methods have been developed using zebra fish larvae, insects, mollusc eggs, algae and higher plants but none has yet demonstrated worthwhile specificity and reproducibility.

In general, biological assays are relatively simple to carry out but do not give absolute specificity and are best used to complement physical and chemical tets. Furthermore, the main bioassays make use of organisms far removed from man on the evolutionary scale. Although mammalian tissue culture bioassays can be used for some mycotoxins they do not yet show high reproducibility. Finally it can be seen that the use of biological methods for the detection of mycotoxins depends on a rapid effect for practical value while it is the long term or chronic response (e.g. carcinogenicity) rather than the acute response of mycotoxins that causes most present concern.

Immunoassay

There has been increasing interest in developing simpler and more specific methods for detecting mycotoxins using immunoassay procedures. Such methods could be of particular significance when dealing with limited sample availability, where extremely low concentrations of mycotoxins are expected, and where large numbers of samples must be analysed. Two techniques have been extensively studied, viz. radioimmunoassay (RIA) and enzyme-linked immunosorbent assay (ELISA).

Immune responses occur when animals form specifically reactive proteins, antibodies, in response to foreign macromolecules. An antigen is any substance which, when introduced into the body, will give rise to the formation of antibodies. Antigens can be proteins, lipoproteins, nucleoproteins, many polysaccharides and most polypeptides. Small molecules, or haptens, can only become antigenic by being linked to proteins or to synthetic polypeptides. Antibodies are highly specific and sensitive compounds capable of detecting trace levels of the specific molecules that they will react with. Mycotoxins are non-antigenic but several mycotoxins have been able to illicit an antibody response in test animals.

Radioimmunoassay (RIA). This method is based on the principle of competition between an unlabelled and a radioactive labelled antigen for a specific antibody present in limited concentrations. The antigen and the specific antibody form a soluble antigen–antibody complex, the process being reversible.

$$\mathrm{Ag} + \mathrm{Ab} \rightleftarrows \mathrm{AgAb}$$

or when labelled

$$\mathrm{Ag}^* + \mathrm{Ab} \rightleftarrows \mathrm{Ag}^*\mathrm{Ab}$$

Thus the Ag* Ab complex will contain the radioactivity of the labelled antigen to the antibody. If unlabelled antigen is now added to the system there will be competition between the unlabelled and labelled antigen for the binding sites of the antibody provided the antibody is in limited concentration.

Thus the higher the concentration of the unlabelled antigen the lower will be the radioactivity of the antigen–antibody complex and the higher that of

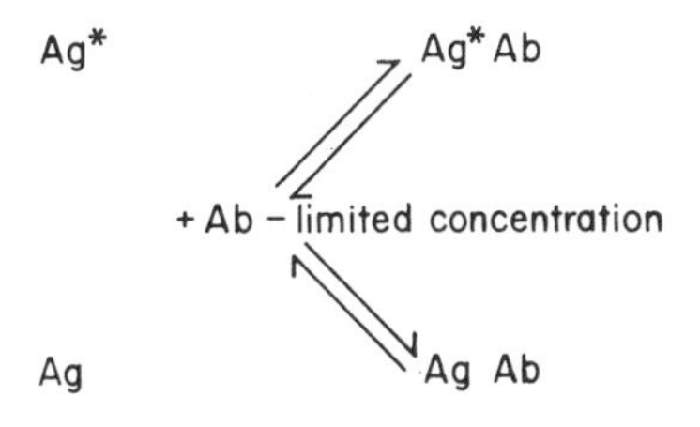

the free labelled antigen. In practice the unlabelled antigen will either be the known standard solution or the unknown sample. The radioactivity of the antibody-bound antigen will decrease while that of the free antigen increases as the concentration of the unlabelled antigen (test sample) increases. Graphical analyses allows accurate determination of the antigen.

The main advantages of the RIA methods are high sensitivity and specificity and the ability to perform simultaneously large numbers of determinations. The method requires antigen in a pure state, specific antisera and a labelled antigen as a tracer. RIA methods have been widely used in medicine and related areas and are now being applied to mycotoxin analysis.

Since mycotoxins are in themselves non-antigenic, protein–mycotoxin conjugates must be developed to immunize animals to produce antibodies specific for the individual mycotoxins. Ochratoxin A can be directly coupled to bovine serum albumin (BSA) in the presence of a water-soluble carbodiimide. Aflatoxin does not have a suitable reactive group and a derivative must be formed. The most successful derivative has been aflatoxin B_1-1-(*o*-carboxymethyl)oxime which has a free carboxyl group suitable for coupling to BSA and subsequent immunization. Similarly, T-2 toxin has been converted to T-2 hemisuccinate and the antigen prepared by conjugation of T-2-HS with BSA.

Immunization schedules and methods of immunization are essentially the same for most RIA systems. The BSA–mycotoxin conjugate in sterile saline solutions emulsified with complete adjuvant (with the object of increasing or stimulating a higher rate of antibody formation in respect of an antigen) is injected into rabbits. Booster injections and bleedings are made at appropriate times after initial immunization and the collected antisera precipitated with $(NH_4)_2SO_4$. Using [^{3}H]-labelled mycotoxin and antisera radio-immunoassays have been developed for several mycotoxins in particular aflatoxins B_1 and M_1, ochratoxin A and T-2 toxin.

With T-2 toxin, studies have shown that the antibody present in the antisera had the greatest affinity for T-2 toxin, less for HT-2 toxin and least for T-2 triol. Weak cross-reaction occurred with 8-acetyl neosolaniol, neosolaniol and no cross-reaction with vomitoxin, trichodermin or diacetoxyscirpenol. Levels of sensitivity were in the range 1–20 μg per assay.

Enzyme-Linked Immunosorbent Assay (ELISA). This immunoassay method is simpler than RIA in not requiring radioactive agents or sophisticated equipment such as a scintillation counter. The ELISA methods are based on the use of antibodies or antigens linked to an insoluble carrier surface. The relevant antigen or antibody in a test solution is then applied and the resulting complex detected by means of an enzyme-linked antibody or antigen. The degradation of the enzyme substrate can normally be measured spectrophotometrically, and is proportional to the concentration of the unknown antigen or antibody in the test solution.

ELISA methods can involve three techniques, viz. detection of antibody using enzyme-labelled anti-globulins, detection of antigen by the double-antibody method or detection of antigen by the labelled antigen competition method. Enzyme markers are usually alkaline phosphatase or peroxidase the reactions of which can be followed spectrophotometrically.

Aflatoxin B_1 and T-2 toxin have been well studied using the detection of antigen by the labelled antigen competition method (Fig. 7.4).

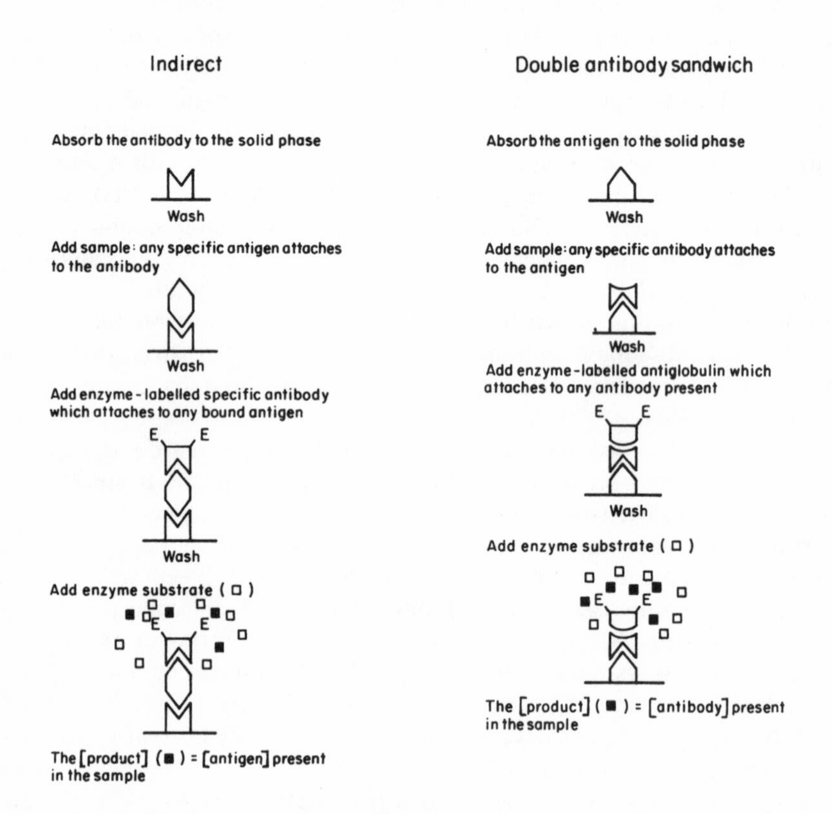

Fig. 7.4. Schematic representation of the ELISA method for antibody determination

In comparative studies between RIA and ELISA methods for analysing mycotoxins ELISA gave more consistent data, relatively lower standard deviations and lower coefficients of variation than did RIA. For aflatoxin B_1 ELISA immunoassays, levels above 5 ppb could be analysed directly after sample extraction without clean-up. This certainly offers a major advantage in time and cost when compared to other methods of quantitation, i.e. TLC, HPLC and GLC. Further advantage of ELISA methods over RIA include shorter assay time (2 h versus overnight) cost per assay, absence of radiation hazards, number of assays that can be dealt with and the possibility of automation.

The greater specificity of monoclonal antibodies could bring further improvements to these immunological assays, and many laboratories are actively pursuing this approach. Successful development of highly specific monoclonal diagnostic kits (now widely available in other fields) for the identification of mycotoxins would have inestimable value.

Safety

In all analytical studies with mycotoxins, strict safety precautions must be practised since many represent a carcinogenic risk to man. Thus it is essential at all times to wear disposable cotton or latex gloves, facial masks and protective clothing. Pure mycotoxins in the dry state, due to their electrostatic nature, will show a tendency to readily disperse in the atmosphere and over work surfaces. This problem can be minimized by making up stock solutions upon receipt of mycotoxins. During the grinding and weighing of contaminated samples there will be the risk of absorbing toxins either through the skin or by inhalation of dust. There can also be the risk of developing allergic reactions due to spores and organic material. In all these cases the risks can be minimized by working under a suitable hood with forced venting and at all times by practising good laboratory hygiene. Care must also be exercised when handling the solvents used for extraction and separation since many are schedule one poisons, e.g. chloroform, acetonitrile (methyl cyanide). When using UV lamps for detecting fluorescent mycotoxins the eyes should be protected by means of UV absorbing glasses.

During analytical procedures equipment such as TLC plates and HPLC columns may retain contamination while surfaces and laboratory space may become contaminated with mycotoxins through unavoidable contact and accidental exposure. Effective methods for the complete destruction of mycotoxins have been developed to ensure safe and environmentally sound disposal practices and these are discussed in Chapter 8 (Castegnaro *et al.*, 1980).

FURTHER READING

AOAC (1975). Chemical hazards of mycotoxins. *Official Methods of Analysis*, 12th edn, chapters 26 and 51. Arlington, USA.

Bennett, G. A., Stubblefield, R. D., Shannon, G. M. and Shotwell, O. L. (1983). Gas chromatograph determinants of deoxynivalenol in wheat. *Journal of the Association of Official Analytical Chemists,* **66**, 1478–1480.

Brown, G. H. (1982). Sampling for 'needles in haystacks'. *Food Technology in Australia,* **34**, 224–227.

Bullerman, L. B. (1979). Methods for detecting mycotoxins in foods and beverages. In: *Food and Beverage Mycology,* L. R. Beuchat(Ed.), Avi Publishing, Connecticut, pp. 445–470.

Castegnaro, M., Hunt, D. C., Sansone, E. B., Schuller, P. L., Sirivardana, M. G., Telling, G. M., Van Egmond, H. P. and Walker, E. A. (1980). *Laboratory Decontamination and Destruction of Aflatoxins B_1, B_2, G_1 and G_2 in Laboratory Wastes,* International Agency for Research on Cancer, IARC Scientific Publications, no. 37, Lyon.

Ceigler, A., Kadis, S. and Ajl, S. J. (Ed.).(1971). *Microbial Toxins, Fungal Toxins,* vol. VI, Academic Press, New York, p. 563.

Davis, W. D., Dickens, J. W., Freil, R. L., Hamilton, P. B., Shotwell, O. L., Wyllie, T. D. and Fulkerson, J. F. (1980). Protocols for surveys, sampling, post-collection, handling and analysis of grain samples involved in mycotoxin problems. *Journal of the Association of Official Analytical Chemists,* **63**, 95–102.

Dickens, J. W. and Whitaker, T. B. (1982). Sampling and sample preparation. In: *Environmental Carcinogens: Selected Methods of Analysis,* vol. 5, *Some Mycotoxins,* International Agency for Research on Cancer, IARC Scientific Publications no. 44, Lyon.

Egan, H., Stoloff, L., O'Neill, I. K., Scott, P., Castegnaro, M., Bertsch, H. (Eds). (1982). *Environmental Carcinogens: Selected Methods of Analysis,* vol. 5, *Some Mycotoxins,* International Agency for Research on Cancer, IARC Scientific Publications no. 44, Lyon.

El-Nakib, O., Pestka, J. J. and Chu, F. S. (1981). Determination of aflatoxin B_1 in corn, wheat and peanut butter by enzyme-linked immunosorbent assay and solid phase radioimmunoassay. *Journal of the Association of Official Analytical Chemists,* **64**, 1077–1082.

Eppley, R. M. (1966). A versatile procedure for assay and preparative separation of aflatoxins from peanut products. *Journal of the Association of Official Analytical Chemists,* **46**, 1218–1223.

Gilbert, J. (1984). The detection and analysis of *Fusarium* mycotoxins. In: *Applied Mycology of Fusarium,* M. O. Moss and J. E. Smith (Eds), Cambridge University Press, Cambridge, pp. 175-193.

Gorst-Allman, C. P. and Steyn, P. S. (1979). Screening methods for the detection of thirteen common mycotoxins. *Journal of Chromatography,* **175**, 325–320.

Haladay, C. E. and Lansden, J. (1975). Rapid screening method for aflatoxin in a number of products. *Journal Agricultural Food Chemistry,* **23**, 1134–1136.

Leaptor, M. S. (1969). Biological assay for aflatoxins. In: *Aflatoxins,* L. A. Goldblatt (Ed.), Academic Press, New York, pp. 107–149.

Moreau, C. (1979). *Moulds, Toxins and Food,* Wiley, New York.

Morgan, M. R. A., McNerney, R. and Chan, H. W-S. (1983). Enzyme-linked immunosorbent assay of ochratoxin A in barley. *Journal of the Association of Analytical Chemists,* **66**, 1481–1484.

Pestka, J. J., Lee, S. C., Lan, H. P. and Chu, F. S. (1981). Enzyme-linked immunosorbent assay for T-2 toxin. *Journal of the Association of Analytical Chemists,* **64**, 940–944.

Pohland, A. E., Thorpe, C. W. and Nesheim, S. (1979). Newer developments in mycotoxin methodology. *Pure and Applied Chemistry,* **52**, 213–223.

Romer, T. R. (1976). Methods of detecting mycotoxins in mixed feeds and feed ingredients. *Feedstuffs,* **12**, 18–21.

Sargeant, K., Carnaghan, R. B. A. and Allcroft, R. (1963). Toxic products in groundnuts: chemistry and origin. *Chemistry and Industry*, 53–55.

Scott, P. M., Kanhere, S. R. and Lan, P.-Y. (1982). Methodology for trichothecenes. *Vth. International IUPAC Symposium on Mycotoxins and Phycotoxins*, Vienna, Austria, 1–3 September.

Shotwell, O. L. and Hesseltine, C. W. (1981). Use of bright greenish-yellow presumptive test for aflatoxin in 1978 corn. *Cereal Chemistry*, **58**, 124–127.

Steyn, P. S. (1981). Multimycotoxin analysis. *Pure and Applied Chemistry*, **53**, 891–902.

Stoloff, L. (1972). Analytical methods for mycotoxins. *Clinical Toxicology*, **4**, 465–494.

Stoloff, L. (Ed.). (1980). *Mycotoxins Methodology*, The Association of Official Analytical Chemists, Washington DC.

Sun, P. J. and Chu, F. S. (1977). A simple solid-phase radio-immunoassay for aflatoxin B_1. *Journal of Food Safety*, **1**, 67–75.

Ueno, Y. (1983). Biological detections of trichothecenes. In: *Trichothecenes: Chemical, Biological and Toxicological Aspects*, Ueno, Y. (Ed.), Elsevier, Amsterdam, pp. 125–133.

Whitaker, T. B. (1977). Sampling granular foodstuffs for aflatoxin. *Pure and Applied Chemistry*, **49**, 1709–1717.

Whitaker, T. B., Dickens, J. W. and Monroe, R. J. (1979). Variability associated with testing corn for aflatoxin. *Journal American Oil Chemists Society*, **57**, 269–272.

Whitaker, T. B., Dickens, J. W., Wiser, E. H. and Monroe, R. J. (1981). Sampling techniques. In: *Food Analysis: Principles and Techniques*, D. W. Gruenwedel and J. R. Whitaker (Eds), Marcel Dekker, New York.

CHAPTER 8

Control of Mycotoxins

The main entry of mycotoxins into the human and animal food chains is from agricultural products such as cereal grains and oil seeds or from products derived from these sources (Chapter 6). From a practical standpoint the best means of restricting mycotoxin contamination is by prevention, in particular, by excluding or reducing toxigenic mould growth in these raw and processed materials.

Prevention can be accomplished by reducing fungal infections in growing crops, by rapid drying and correct storage of the harvested crops, or by using effective anti-mould preservatives. When mycotoxin contaminated material is suspected or identified it may in part be salvaged by the removal of contaminated material by mechanical separation techniques, by chemical extraction of the mycotoxins or by decontamination or detoxification of the material by physical, chemical or biological techniques. The use of each method will be dependent on the type of contaminated material and the use to which the material will be directed. Financial, nutritional and toxicological considerations will temper these decisions.

PREVENTION

Mycotoxins can be produced by toxigenic moulds growing on (1) living plants, (2) decaying plant material and (3) stored products. Toxigenic mould spores are almost universally present in the atmosphere but particularly with agricultural crops and products, and successful containment programmes will involve the development of methods to inhibit the germination and proliferation of these spores. High moisture levels (20–25 per cent wet weight) are required for the growth of fungi in the living plant or in decaying organic material whereas fungi capable of growing in stored products can usually exist on moisture contents of 12–18 per cent.

Pre-harvest treatments

Many mycotoxins, in particular the *Fusarium* mycotoxins zearalenone and the trichothecenes, the ergot alkaloids, the tremorgens and, more recently observed, the aflatoxins, can be formed during the growing stages of certain crops. Many environmental conditions have been identified which can promote mycotoxin formation in growing crops and include insect infestation, drought conditions, varietal susceptibility or resistance, mechanical damage, nutritional deficiencies and unseasonal temperatures and rainfall.

On-farm preventative techniques will largely revolve around good farm management, in particular, methods of cultivating to improve plant vigour, the judicious use of insecticides and fungicides to reduce insect and fungal infestation, irrigation to avoid drought conditions and, more recently, genetical methods. Attempts are now being made to develop commercially acceptable varieties of crops, e.g. peanuts, maize, cottonseed, that will be more resistant to toxigenic moulds or will inhibit toxin production. For example, the 'hard seed coat' trait in cottonseed has been shown to be more resistant to *Aspergillus flavus* penetration and similar lines of research are now being carried out with peanuts.

Post-harvest treatments

Post-harvest contamination with mycotoxins can be controlled only through the use of current technology and knowledge derived from a broad spectrum of scientific disciplines—physiology, biochemistry, microbiology, agriculture and engineering (Lieberman, 1983). Post-harvest technology is expensive but essential if the gains made by improving world agriculture are not to be dissipated by post-harvest deterioration. For example, losses of grain annually amount to about 10–20 per cent of world production (Williams, 1983). The handling of agricultural produce within communities and on a world export basis is a formidable task (Table 8.1).

Table 8.1. Components of a grain transportation system (Williams, 1983)

1. Collection of grains from farms into consolidated deposits
2. Facilities for storage, short- and long-term
3. Loading, unloading and conveying systems
4. Methods of packaging or bulk handling
5. Roads, railways and waterways
6. Systems for grading the commodities into categories of different visual quality and establishing equitable price scales
7. Systems for servicing and maintaining equipment and facilities for all aspects of the system
8. Systems for recruiting and training personnel for operation and administration
9. Systems for education and extension of information to farmers, grain merchants and other personnel involved with the overall handling operation

Physical methods

In order to prevent or to delay mould spoilage of foods and feeds different physical methods of control have been practised, e.g. heating, pasteurization and sterilization, cooling (cold or frozen storage), vacuum packing, canning, drying and irradiation. In some instances, the organoleptic qualities of a food suffers greatly from these physical modes of preservation while in many instances these methods are not technically or economically feasible. The choice of method will be very dependent on the scale of operation. Since the greatest potential for mould growth and mycotoxin formation will occur early in the food or feed chain, particular attention will be given to treatments of raw agricultural products.

After harvest and during shipment, storage and compounding of agricultural commodities, toxigenic mould growth and subsequent mycotoxin production (particularly with the toxins produced by *Aspergillus* and *Penicillium* spp.) will be influenced by moisture levels, temperature, aeration, mould spore density, conditions of storage, in particular, leakage of water or condensation, biological heat and, of course, the chemical and nutritive nature of the material (Table 8.2). By far the most critical environmental factors determining whether a substrate will support mould growth are temperature, moisture content and time. Each of these parameters can be used creatively for prevention of mould growth.

Table 8.2. Measures for prevention of damage to grains during storage and transportation (Williams, 1983)

Aeration	Storage bins, short-term and long-term
Moisture control	Storage bins and containers
Temperature control	Storage bins and warehouses
Fumigation	Storage bins, warehouses, ships' holds, railcars, trucks, sacks
Fungicides	Storage bins, warehouses
Insecticides	Storage bins, warehouses, ships' holds, railcars, trucks, sacks
Design	Storage areas
Hygiene	All areas
Legislation	All areas

Low temperature storage can be very suitable for controlling mould growth, and is an essential feature of many commercial storage programmes. The use of cold temperatures for large-scale storage of agricultural crops is generally not economically feasible. Water activity (a_w) is a parameter being used instead of moisture content and expresses the availability of water for the growth of microorganisms. The lower the a_w the less available is the substrate for microorganisms. Fungi generally have much lower minimal a_w values than bacteria and this explains why many products free of bacterial spoilage can be spoiled by fungi. In practice fungal growth can be prevented by drying products to a level below a_w 0.65 and retaining the material at that level. Somewhat similar effects can be achieved by changing the osmotic level of a material either by the addition of salts or sugar (Carey, 1978). Finally, the time between harvesting and preventative treatments as well as the length of time that a material must await utilization (shelf-life) are of critical importance when it is realized that mould growth and mycotoxin formation can occur in relatively short time scales (e.g. 2–3 days).

Agricultural products, in particular, cereals and oil seeds, will normally be held under storage for varying periods of time. The prevention of mould growth in storage conditions is normally accomplished by the modification of the inter-seed environment, e.g. controlling moisture, temperature and the gaseous atmosphere. In most cases the control of moisture content is the main commercial approach used to inhibit fungal growth and subsequent mycotoxin formation. Under storage conditions insects can be extremely supportive to mould growth since with heavy infestations they can increase moisture levels and temperature by their metabolic activity, destroy the protective testa of seeds and, furthermore, can carry spores from pockets of infection throughout the total stored commodity. For these reasons, insect infestation must always be rigorously dealt with.

The achievement of long-term storage is now recognized as both a microbiological and an engineering problem. Initially, a drying regime will reduce the moisture level below the value required for mould spore germination. The dried material will then be placed under storage conditions where environmental control by engineering practices will attempt to hold the moisture level at the minimal value (Fig. 8.1). Within large volumes of stored materials changing environmental conditions or physical damage to the containment vessel, e.g. storage bin, allowing the entry of water, can lead to migration of moisture, condensation and formation of moisture pockets. In these isolated conditions mould spore germination, proliferation and toxin production can occur. When the stored material is then withdrawn there may well be a lack of uniformity in the material with irregular occurrence of mycotoxins.

In practice, it has been shown that the greatest risks will occur with the storage of inadequately dried agricultural products and rewetting of dried and stored products. On-farm storage of cereals and animal feeds is particularly problematic in countries with high ambient humidity.

Throughout the world many different methods have been developed and applied to ensure rapid drying of agricultural products. High temperature/high speed drying is practised in USA for high moisture maize. While this method is beneficial in reducing moisture levels to 19 per cent, going below this level can often lead to excessive seed cracking. Slow drying methods make use of natural or ambient air drying in which air is forced through a grain bulk. This method produces excellent results and has become an alternative to the ancient methods of repeated 'turning' of grain masses. Low temperature drying can be obtained by increasing air temperature by 1.5–5.5° C to dry grain in the bin more quickly and can be independent of weather. Solar drying has vast potential and is similar to low heat drying. The air temperature can be increased as much as 16–17° C.

Fig. 8.1. Grain storage silos

Grain and other agricultural produce may be stored under hermetic or anaerobic conditions in silos. Initially the raw material will undergo fermentation by surface microorganisms which will ensure rapid depletion of oxygen and build up of carbon dioxide. Most fungi are aerobic and cannot grow or produce mycotoxins in the absence of oxygen. Although most silos, if properly managed, will create truly anaerobic conditions, improper management will result in a partial or total aerobic atmosphere being generated inside the silo. When this occurs moulds can flourish and mycotoxins may be formed. Instances of patulin occurrence are well-documented. Even under ideal practice once the ensiled grains are removed from storage they are highly susceptible to mould attack and toxin production. Although this method is energy efficient, it does restrict management flexibility since the grain must be used for feed purposes at or near storage.

The duration of storage is obviously an important factor when considering mycotoxin formation. The longer the stay in storage bins, the greater will be the possibility of changes in environmental conditions conducive to mould proliferation. Large-scale commercial operation gives due attention to retaining correct environmental programmes during long-term storage. However, in smaller on-farm operations, bin hygiene seldom rates high in farm management, and months, even years, may pass with no routine clean-out.

Chemical methods

Where physical methods of preservation are not satisfactory or cannot easily be applied, the use of chemical preservatives becomes necessary. Mould growth on stored crops can be reduced or prevented with small capital outlay by the addition of selected chemicals at harvest to decrease the need to achieve or retain low available water contents. Similar use can occur with processed foods and feeds and several preservatives suitable for different classes of materials are now routinely used.

In practice, preservatives are substances which inhibit or kill microorganisms in food and feeds. International regulations vary somewhat in the types and levels of preservatives that are acceptable. Table 8.3 shows the Acceptable Daily Intake (ADI) permitted for preservatives by the FAO/WHO Expert Committee on Food Additives. A successful preservative must have very low mammalian toxicity but possess wide and long lasting microbial inhibitory properties. Preservatives will normally not reduce the concentration of mycotoxins present in food and feeds before the preservative was added. The mechanism of action of the preservatives appears to be mainly by enzyme inhibition in the cell (sorbic acid, propionic acid) or by the destruction of the cell membrane (natamycin). At the prescribed preservative concentrations they are mainly fungistatic resulting in inhibition but not killing of the fungi.

Table 8.3. Acceptable daily intake (ADI) of preservatives

	ADI (mg kg^{-1} body weight per day)
Propionic acid	no limit
Sorbic acid	0–25
p-Hydroxybenzoic acid esters	0–10
Nitrate	0–5
Benzoic acid	0–5
Formic acid	0–3
Sulphur dioxide, sulphites	0–0.7
Natamycin	0–0.3
Nitrite	0–0.2
Hexamethylene tetramine	0–0.15

Various organic acids such as sorbic, benzoic, propionic, acetic and formic acids have been widely used as preservatives, particularly of stored agricultural products. They are normally used as the corresponding sodium, potassium or calcium salts which are more water soluble. Sorbic, benzoic and propionic acids exert their antimicrobial activity only when they are present as undissociated acids, functioning more efficiently at lower pH values (Table 8.4).

Propionic acid alone or in combination with sorbic acid and acetic acids has been increasingly used to control mould growth in stored agricultural products, particularly allowing the preservation of high moisture maize and animal feeds. Propionic acid functions largely as an anti-germination agent

Table 8.4. Percentage undissociated acid at different pH values

	Percentage undissociated acid at pH				
Preservative	3.0	4.0	5.0	6.0	7.0
Formic acid	85	36	5	0.6	0.06
Benzoic acid	94	61	13	1.5	0.15
Acetic acid	98	85	36	5.4	0.6
Sorbic acid	98	85	37	5.5	0.6
Propionic acid	99	88	43	7.0	0.8

but if germination does occur the acid can be readily metabolized by most fungi. Successful chemical preservation of agricultural material requires adequate distribution of a suitable dose of the chemical throughout the mass. With volatile preservatives like propionic acid up to 70 per cent may be lost during application. The level of water in the crop must be accurately determined in order to predict the correct dose which must be retained on the stored material. Uneven distribution can allow moulds to grow throughout a treated material from small initial foci and can modify the sequence of moulding to encourage growth of single species or restricted groups of species and so perhaps increase the possibility of mycotoxin formation (Al Hilli and Smith, 1979).

Propionic acid has a number of disadvantages as a grain preservative, viz. it is volatile, a condition which aids distribution but also contributes to application losses, it has a pungent odour and its acidic corrosive nature makes it unpleasant and hazardous to handle in concentrated form.

Sorbic acid and natamycin have been shown to be especially effective against toxigenic fungi and are used in Europe to inhibit toxigenic moulds on cheeses and dry sausages. Sorbic acid is widely used as preservative in bakery products. Diphenyl, *O*-phenylphenol and thiabendazole are widely used to inhibit mould growth on citrus fruit, used either as a fruit wash or impregnated in wrapping paper. Gentian violet, historically of major importance in medical mycology, is now being increasingly used as a preservative in the animal feed industry.

A new class of mould inhibitor (N-oxime ethers) has been shown to be equal to or superior to propionic acid in controlling moulds but since these inhibitors are heat labile the range of use will be limited.

ELIMINATION OF MYCOTOXINS FROM PRODUCTS

Once a product is contaminated with mycotoxins there are only two options if it is to be used for human or animal consumption:

1. The toxin can be removed;
2. The toxin can be degraded into less toxic or non-toxic compounds.

How inherently stable are mycotoxins in biological systems? The most studied mycotoxin is aflatoxin which appears to be quite stable under normal biological conditions. Patulin and penicillic acid are, however, relatively unstable in some food products due to the combination of these mycotoxins with amino acids and sulphydryl groups. Ochratoxin A is less stable than aflatoxin but more stable than patulin or penicillic acid. However, there is need for more studies to assess the long-term stability of the more common mycotoxins (Chapter 6) regularly found in foods and feeds.

Physical separation

In large particle size agricultural products such as the peanut, Brazil nut or almonds, aflatoxin contamination when it occurs is normally confined in any batch to a small number of contaminated seeds or kernels. It has been shown that the level of aflatoxin in peanuts can be correlated with the proportion of loose-shelled kernels and the number of shrivelled, rancid or discoloured kernels. When these are discarded the remaining kernels are relatively free of aflatoxin. Off-coloured kernels normally imply mould contamination and such kernels can now be separated either by hand or by passing through colour sorters. Colour sorters are electronic devices which can identify the infected samples and divert them from the main stream by a jet of air. Discarded kernels can be used for oil stock and other non-edible uses. New bi-chromatic colour sorters can now distinguish various shades of colour as well as intensity, e.g. yellow, red or green tints. Aflatoxin contamination of Brazil nuts and almonds can also be significantly reduced by pneumatic sorting ensuring a high standard of quality for these products in the confectionary industry.

Chemical separation

Numerous processes have been developed to remove aflatoxin from contaminated materials (particularly peanuts) by various chemical extraction techniques. Several procedures can be used to remove aflatoxins from oilseeds and meals and include extraction of aflatoxins with appropriate solvents, simultaneous solvent extraction of oil and aflatoxin and selective extraction of aflatoxins from peanut oil with sodium hydroxide and bleaching earth. Although solvent extraction can be highly successful the cost of the additional processing, the need for special solvent-removing equipment, loss of nutrients from the residual meals, etc., have made these processes of questionable economic value. Material treated in this way would only be suitable for animal feed. Chemical extraction of other common mycotoxins has not yet achieved practical acceptance.

Degradation or detoxification

Many physical, chemical and biological methods have been investigated to degrade *in situ* mycotoxins present in raw material. Almost all studies have been concerned with aflatoxins (Tables 8.5 and 8.6).

Physical methods

Irradiation. Aflatoxins are sensitive to ultraviolet light. The degradation of aflatoxin in contaminated products is dependent on the nature of the solvent, toxin concentration and the length of exposure to the UV light. As yet there is no clear indication that UV treatment offers a worthwhile approach for practical purposes.

Heat. Aflatoxins, together with many other mycotoxins, are relatively heat-stable. With dry heat such as roasting, temperatures approaching the melting point (250°C) of aflatoxin must be used to effect any degradation of the toxin. In the presence of moisture much lower temperatures can cause a marked degradation of aflatoxin. Increasing moisture content and/or time of heating increases the rate of aflatoxin degradation. However, the adverse effects of heat treatment on the appearance and nutritive value of the product makes the practical application of these methods highly doubtful.

Chemical. A wide range of chemicals have been tested as reagents for the destruction of aflatoxins including acids, alkalis, aldehydes, oxidizing agents and several gases.

Although strong acids can effectively degrade aflatoxins, it is unlikely that they would be used practically to detoxify agricultural commodities since these methods are too drastic and would change the properties of the product. Several relatively inexpensive methods for decontaminating agricultural commodities make use of inorganic or organic bases. Thus NaOH can be used for destroying aflatoxin in refining oil while $Ca(OH)_2$ and formaldehyde have been used to reduce aflatoxin in peanut meal.

Ammonia used as an anhydrous gas at elevated temperatures and pressures can cause a 95–98 per cent reduction in total aflatoxin concentration in peanut meal. It is believed that the ammonia opens the lactone-ring of aflatoxin B_1, forming an ammonium salt of the resulting hydroxyacid. Since the reaction is carried out at elevated temperature and pressure, it causes the decarboxylation of the β-keto acid to the so-called aflatoxin D_1 (Fig. 8.2). Ammonia treatment of aflatoxin-contaminated corn flour resulted in the covalent binding of aflatoxin B_1 to corn protein and water soluble constituents of the corn. This method is now being used on a commercial scale in USA and although the protein efficiency rate values of the product are lowered, the chemical composition of the meals altered, plus some degree of off-flavours and off-odours, the final products can still be used for animal feed.

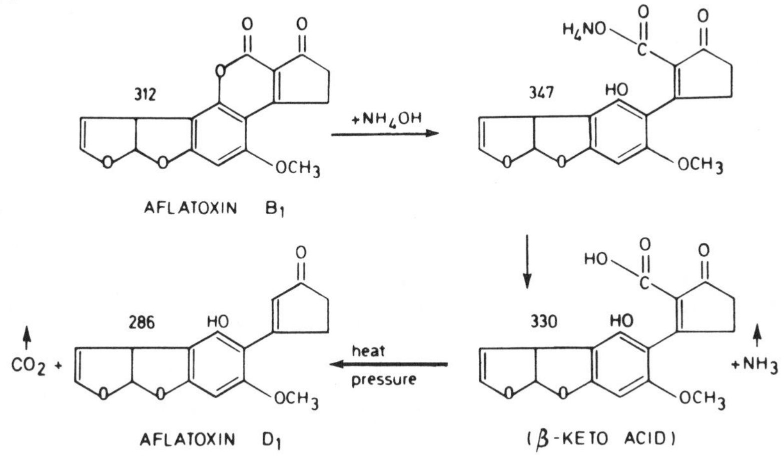

Fig. 8.2. Proposed scheme for formation of the major products of ammoniation of aflatoxin B_1

Oxidizing agents such as hydrogen peroxide have shown considerable promise for decontaminating aflatoxin-contaminated foods and feeds. Similarly, bisulphite has been shown to degrade aflatoxin B_1 in naturally contaminated maize and to be cost competitive with the ammoniation process. However, limitations associated with the decontamination process (time and temperature) and the insufficient degradation of aflatoxin B_2 will make this process untenable.

Biological. The natural ability of microorganisms to degrade most molecules has been a source of investigation for the biological breakdown of mycotoxins. Many microorganisms including bacteria, actinomycetes, yeasts, moulds and algae show varying abilities to degrade aflatoxin. The most active organism so far discovered is *Flavobacterium aurantiacum* which in aqueous solution can take up and metabolize aflatoxins B_1, G_1 and M_1. As yet no commercial application has been developed. In co-cultivation experiments with toxigenic and non-toxigenic moulds there are many instances where marked reductions in toxin presence has been observed. Whether this is due to inhibition of production or breakdown of the toxin has not yet been determined. Correspondingly, stimulation of toxin production under such conditions has also been observed.

Table 8.5. Selection of methods which have been used to degrade aflatoxins in various media and environments (Castegnaro *et al.*, 1980)

Medium	Aflatoxins checked for degradation	Degradation method	Efficiency
Peanut meal	B_1	Treatment with excess anhydrous ammonia gas (10–30 g per kg of meal) at elevated temperature and pressure for 30 to 60 min	97%
Peanut meal	B_1-B_2-G_1-G_2	Combined action of formaldehyde and calcium hydroxide at 116 °C in presence of 20% moisture	Aflatoxin content decreases to below 20 $\mu g\ kg^{-1}$
Cottonseed meal	Aflatoxins	Ammoniation at 10% moisture, under various conditions; best results are obtained with 4% ammonia after 65 min	98%
Corn	B_1 and B_2	Treatment with low level of ammonia (1.5%) at low ambient temperature (– 11 °C to + 16 °C) for 179 days	Better than 96%
Peanut and cottonseed meal	B_1-B_2-G_1-G_2	A large variety of acids, anhydrides, inorganic bases, oxidizing agents nitrogen organic compounds and aldehydes have been tested on a laboratory scale to detoxify aflatoxin-contaminated peanut and cottonseed meals. Reaction conditions using methylamine, sodium hydroxide and formaldehyde had been developed	

Oilseed meal	B_1-G_2	Treatment with finely divided calcium hydroxide (50 μm mean particle size) at 80 °C for 10 min	100%
Crude oil from peanut or corn	B_1	Use of normal refining process and treatment with NaOH and bleaching earth	Final concentrations below level of detection
Solutions	B_1-B_2-G_1-G_2	Treatment with any of the following reagents: NaOCl (6.8 M), $KMnO_4$ (0.63 M), $Ce(NH_4)_2(SO_4)_3$ (0.63 M), Na_2CO_3 (0.34 M)	Complete degradation
TLC plates		Treatment with any of the following reagents: NaOCl, $KMnO_4$, phenol, chlorohydroquinone, resorcinol, $Na_2S_2O_3$, $NaBO_3$, $NaBO_2$, HCl, NaOH or combination with H_2O_2	Rapid degradation
Work area surfaces		Wipe surface with a cloth impregnated with dilute NaOCl solution	
Personnel		Hand: 5% to 6% solution NaOCl Mouth: gargle with 1% perborate and 1% sodium bicarbonate	Reduces aflatoxins below the level of detection
Disposable commodities		Soak in undiluted bleach for 30 min. Mechanical mixing may be required to ensure complete contact with the bleach	

Table 8.6. Chemical reactions of aflatoxins and toxicity of the reaction products (Castegnaro *et al.*, 1980)

Aflatoxin	End product	Reaction and % conversion	Toxicity of end product
All aflatoxins		Dilute solutions of NaOC1 destroy both the aflatoxin-producing mould and the aflatoxins. Efficiency, 100%	
B_1	B_2	Treat with cold, aqueous, mineral acid	Test on 1-day old ducklings
G_1	G_2		No significant difference in growth between control and dosed birds and no liver lesions associated with aflatoxin poisoning
B_1	B_2	Treat with citric acid solution at 28 °C for 24–48 hours	In duckling test, 55 μg of end product caused no deaths compared with an LD_{50} of 40 μg of aflatoxin B_1
B_1	Unidentified	Treat with ammonia at room temperature. Aflatoxin may be regenerated by acidification of the unidentified product	Non-toxic to chick embryo
B_1	Succinic acid	Treat with 1 mol N NaOH and 30% H_2O_2 for 70 min.	
B_1	Three compounds:	Irradiate with UV light in methanolic solution.	

	(I) 1-hydro-2-6-dimethoxyfurocoumarone (II and III) 2 isomers of 2-hydro-1-6-dimethoxyfurocoumarone	Conversion, 92% (24% of I, 30% of II, 38% of III)	
B_1-G_1	Two major degradation products are isolated per aflatoxin	Irradiate with UV light (long wave) for 1 h	No toxicity shown on chick embryo at the level tested
B_1-G_1		Irradiate with UV light	The degradation products are toxic, although less so than aflatoxin
B_1 G_1	Probably D_1	Autoclave groundnut meal containing 60% moisture at 15 lb in^{-2} (120 °C) for 4 h. Conversion of B_1, 95%	Very slight toxicity was shown on duckling tests
B_1		Baking at 120 °C for 30 min. does not affect the content of added aflatoxin in bread	
B_1		Clay is very efficient in removing aflatoxin from solutions. Desorption is then very difficult. Adsorption normally 90–100%	

Of considerable current interest is the demonstration that older mycelia of toxigenic fungi can break down the producer toxin. Intramycelial protein extracted from a fragmented mycelium of *A. parasiticus* is capable of degrading aflatoxin. The reaction was further activated by addition of hydrogen peroxide. The possibility of producing mycotoxin degrading enzymes would have considerable commercial interest and application.

In conclusion, the prevention of mycotoxin formation in agricultural produce and other foodstuffs represents both a pre- and post-harvest problem of regulating the environmental factors influencing fungal growth. The grower must practise good farm management, in particular, controlling insects and fungal pests, harvesting with minimal damage to seed coats, and the use of adequate drying techniques to rapidly reduce seed moistures to safe levels for storage. Food processing should aim to exclude moulds at all stages of production and distribution and should further analyse raw materials and products for the presence of mycotoxins. The growing international awareness of the dangers of mycotoxins to the health of man and animals is now being more realistically appreciated. Strict adherence to the multidisciplinary principles of mycotoxin control (Fig. 8.3) would dramatically reduce the exposure of human and animal populations to the deleterious effects of mycotoxins.

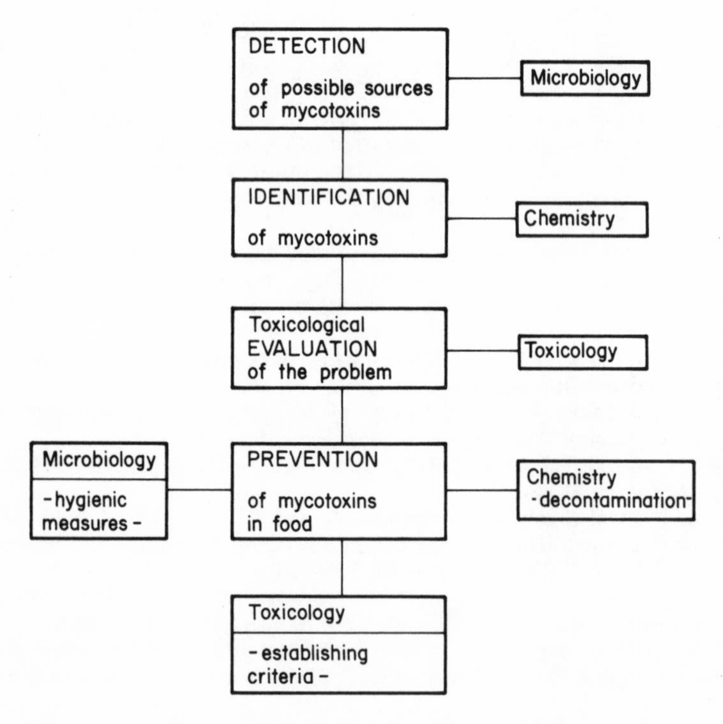

Fig. 8.3. Multidisciplinary approach of mycotoxin research (Samson *et al.*, 1981)

FURTHER READING

Al-Hilli, A. L. and Smith, J. E. (1979). Influence of propionic acid on growth and aflatoxin production by *Aspergillus flavus. FEMS Microbiology Letters*, **6**, 367–370.

Beckwith, A. C., Vesonder, R. F. and Ciegler, A. (1975). Action of weak bases upon aflatoxin B_1 in contact with macromolecular reagents. *Journal of Agricultural Food Chemistry*, **23**, 582–588.

Bennett, G. A., Shotwell, O. L. and Hesseltine, C. W. (1980). Destruction of zearalenone in contaminated corn. *Journal American Oil Chemists' Society*, **56**, 812–819.

Bullerman, L. B. (1979). Significance of mycotoxins to food safety and human health. *Journal of Food Protection*, **11**, 65–86.

Burditt, S. J. and Hamilton, P. B. (1983). Mold inhibitory activity of an *N*-oxide ether. *Poultry Science*, **62**, 2183–2186.

Castegnaro, M., Hunt, D. C., Sansone, E. B., Schuller, P. L., Siriwardana, M. G., Telling, G. M., Van Egmond, H. P. and Walker, E. A. (1980). *Laboratory Decontamination and Destruction of Aflatoxins* B_1, B_2, G_1 *and* G_2 *in Laboratory Wastes*, International Agency for Research on Cancer, IRAC Scientific Publications, Lyon.

Christensen, C. M. (Ed.) (1974). *Storage of Cereal Grains and their Products.* The American Association of Cereal Chemists, St. Paul, Minnesota.

Christensen, C. M. and Kauffman, H. H. (1969). *Grain Storage*, University of Minnesota Press, Minneapolis, Minnesota.

Corry, J. E. L. (1978). Relationships of water activity to fungal growth. In: *Food and Beverage Mycology*, L. R. Beuchat (Ed.), Avi Publishing, Connecticut, pp. 45–82.

Cucullin, A. F., Lee, L. S., Pans, W. A. Jr. and Stanley, J. B. (1976). Ammoniation of aflatoxin B_1. *Journal of Agricultural Food Chemistry*, **24**, 408–414.

Goldblatt, L. A. and Dollear, F. G. (1977). Review of prevention, elimination and detoxification of aflatoxins. *Pure and Applied Chemistry*, **49**, 1759–1764.

Hagler, W. M. Jr., Hutchins, J. E. and Hamilton, P. B. (1982). Destruction of aflatoxin in corn with sodium bisulfite. *Journal of Food Protection*, **45**, 1287–1291.

Hall, D. W. (1970). Handling and storage of food grains in tropical and subtropical areas. F.A.O. Agricultural Development Paper no. 90, Rome, Italy.

Huitson, J. J. (1968). Cereals preservation with propionic acid. *Process Biochemistry*, **3**, 31–32.

Lieberman, M. (ed). (1983). *Post-Harvest Physiology and Crop Preservation*, NATO Advanced Study Institutes Series, series A, vol. 46.

Lord, K. A. and Lacey, J. (1978). Chemicals to prevent the moulding of hay and other crops. *Journal of Science Food and Agriculture*, **29**, 574–575.

Marth, E. H. and Doyle, M. P. (1979). Update on molds: degradation of aflatoxin. *Food Technology*, 81–86.

Maerck, K. E., McElfresh, P., Wohlman, A. and Hilton, B. W. (1980). Aflatoxin destruction in corn using sodium bisulfite, sodium hydroxide and aqueous ammonia. *Journal of Food Protection*, **43**, 571–574.

Sampson, R. A., Hoekstra, E. S. and Van Oorschot, C. A. N. (1981). *Introduction to Food-Borne Fungi*, Centraal bureau voor Schimmelcultures, Delft.

Scott, P. M. (1973). Mycotoxins in stored grains, feeds and other cereal products. In: *Grain Storage—Part of a System*, R. V. Sinha and W. E. Muir (Ed.), Avi Publishing, USA.

Tuite, J. and Foster, G. H. (1979). Control of storage diseases of grain. *Annual Review of Phytopathology*, **17**, 343–360.

Williams, P. C. (1983). Maintaining nutritional and processing quality in grain crops during handling, storage and transportation. In: *Post-Harvest Physiology and Crop Preservation*, M. Lieberman (Ed.), NATO Advanced Study Institutes Series, series A, vol. 46, pp. 425–444.

Index